Flow Cytometry Applications in Cell Culture

edited by

Mohamed Al-Rubeai
A. Nicholas Emery

School of Chemical Engineering
The University of Birmingham
Birmingham, England

CRC Press
Taylor & Francis Group
Boca Raton London New York

CRC Press is an imprint of the
Taylor & Francis Group, an **informa** business

CRC Press
Taylor & Francis Group
6000 Broken Sound Parkway NW, Suite 300
Boca Raton, FL 33487-2742

First issued in paperback 2019

© 1996 by Taylor & Francis Group, LLC
CRC Press is an imprint of Taylor & Francis Group, an Informa business

ISBN-13: 978-0-8247-9614-3 (hbk)
ISBN-13: 978-0-367-40149-8 (pbk)

Library of Congress Cataloging-in-Publication Data

Flow cytometry applications in cell culture / edited by Mohamed Al-Rubeai, A. Nicholas Emery.
    p.   cm.
    Includes index.
    ISBN: 0-8247-9614-4 (hbk. : alk. paper)
    1. Flow cytometry. 2. Cell culture. I. Al-Rubeai, Mohamed, II. Emery, A. Nicholas.
QH585.5.F56F574 1996
574.87'0724—dc20
                        95-40326
                        CIP

Visit the Taylor & Francis Web site at
http://www.taylorandfrancis.com

and the CRC Press Web site at
http://www.crcpress.com

Preface

Flow cytometry is now a standard technique, with many applications in the biological and medical sciences, and much effort has been applied to advancing the instrumentation and techniques in both research and diagnostic fields. More recently this has led to growing interest in its use for the identification, characterization, monitoring, and control of microbial, plant, and animal cell bioprocesses. It is now possible to obtain information about cell populations unobtainable by other means and of a form and quality far superior to that available by conventional biochemical analysis and that can be available in a near on-line mode and timescale. With such information readily available, it should be possible to make significant progress in the optimization of large-scale cell culture for the successful manufacture of those many valuable products that are arising as a direct result of recent developments in molecular and cell biology, and even to consider the flow cytometer as a near on-line process sensor for control purposes. In particular, the increasing availability of fluorescing markers for the analysis of cell structure and function is likely to make significant contributions to the elucidation of the population dynamics and hence the development of sophisticated models of cell growth and product synthesis and secretion.

This book brings together the knowledge and experience of those who are applying flow cytometry to the investigation and monitoring of cellular processes for biotechnological purposes. Chapters review the state of the art with in-depth assessments that emphasize the practical aspects of efficient operation of flow cytometric techniques. The book indicates those aspects of cell variability encountered in cell culture processes and shows how the quantitative analysis of heterogeneous populations can help full realization of the potential of biotechnological exploitation of cells. Cell cycle analysis of animal cell cultures is of particular importance as an aid to the early estimation of process state and to the prediction of future culture performance. Many of the authors lay emphasis on such studies. The importance of cell line selection and product stability are now widely recognized, and information is presented here on the use of flow cytometry to monitor and improve the selection and enrichment of high-productivity cell lines. Analysis of baculovirus-infected insect cell culture is dealt with as are a number of flow cytometry techniques of varying complexity that have been developed to detect cellular and subcellelular changes occurring during apoptosis.

Flow cytometry offers a rapid and versatile means to analyze the properties of hematopoietic cells. Its use to characterize ex vivo expanded human bone marrow is presented with detailed protocols. The applications of flow cytometry to studies of microbial populations are also described, including the determination of population structure and cell cycle progression in yeast and the analysis of bacterial growth and viability. The potential for flow cytometry to characterize plant cell populations together with the problems associated with analysis and sorting of plant cells are included to provide a sound basis for future developments in this newly opened area.

Individual chapters provide reviews by experienced practitioners of the specific methodologies available together with outlines of recent data obtained and insights gained. The material is addressed primarily to those interested in flow cytometry in microbial and cell culture processes, but many topics will also be of interest to microbiologists, biochemists, and other bioscientists as well as to biochemical engineers. The editors come from bioscience (Al-Rubeai) and engineering (Emery) backgrounds and are therefore well placed to ensure that the contents are addressed to and understandable by a wide range of disciplines. We are enthusiasts for the cause of flow cytometry—we hope that our readers find inspiration to explore further its potential themselves and that this work helps their rapid progress.

Mohamed Al-Rubeai
A. Nicholas Emery

Contents

Contributors

Mohamed Al-Rubeai School of Chemical Engineering, The University of Birmingham, Birmingham, England

Lilia Alberghina Dipartimento di Fisiologia e Biochimica Generali, Sezione Biochimica Comparata, Università degli Studi di Milano, Milan, Italy

R. Andrew Badley Immunology Department, Colworth Laboratory, Unilever Research, Sharnbrook, Bedford, England

Georges Belfort Department of Chemical Engineering, Rensselaer Polytechnic Institute, Troy, New York

Rolf Bernander Department of Biophysics, Institute for Cancer Research, Oslo, Norway

E. Coen Beuvery National Institute for Public Health and Environmental Protection (RIVM), Bilthoven, The Netherlands

Erik Boye Department of Biophysics, Institute for Cancer Research, Oslo, Norway

David Brott Aastrom Biosciences, Inc., Ann Arbor, Michigan

Marc Cherlet Laboratoire des Sciences du Génie Chimique—CNRS, Institut National Polytechnique de Lorraine, Vandœuvre-lès-Nancy, France

Florence K. F. Chua Bioprocessing Technology Centre, National University of Singapore, Singapore

José M. Coco Martin Division of Experimental Therapy, The Netherlands Cancer Institute, Amsterdam, The Netherlands

John F. Dunne Syntex Research, Palo Alto, California

Clive Edwards Department of Genetics and Microbiology, University of Liverpool, Liverpool, England

A. Nicholas Emery School of Chemical Engineering, The University of Birmingham, Birmingham, England

Jean-Marc Engasser Laboratoire des Sciences du Génie Chimique—CNRS, Institut National Polytechnique de Lorraine, Vandœuvre-lès-Nancy, France

Patricia Franck Centre Hospitalier Régional Universitaire de Nancy, Nancy, France

David W. Galbraith Department of Plant Sciences, University of Arizona, Tucson, Arizona

Guenter Giese Max-Planck Institute of Cell Biology, Ladenburg, Germany

Bernhard Goller Department of Cell Biology and Biotechnology, Boehringer Mannheim GmbH, Penzberg, Germany

Forest Gray Syntex Research, Palo Alto, California

Christopher D. Gregory Department of Immunology, The University of Birmingham Medical School, Birmingham, England

Anthanassia K. Kioukia Radioimmunochemistry Laboratory, "Demokritos" National Centre for Scientific Research, Athens, Greece

Manfred R. Koller Aastrom Biosciences, Inc., Ann Arbor, Michigan

Manfred Kubbies Department of Cell Biology and Biotechnology, Boehringer Mannheim GmbH, Penzberg, Germany

Georgina M. Lambert Department of Plant Sciences, University of Arizona, Tucson, Arizona

Kristina G. Lazzari One Cell Systems, Inc., Cambridge, Massachusetts

V. Leelavatcharamas School of Chemical Engineering, The University of Birmingham, Birmingham, England

James B. Lin One Cell Systems, Inc., Cambridge, Massachusetts

Annie Marc Laboratoire des Sciences du Génie Chimique—CNRS, Institut National Polytechnique de Lorraine, Vandœuvre-lès-Nancy, France

Katherine L. McKinney[*] Department of Chemical Engineering, Rensselaer Polytechnic Institute, Troy, New York

Anne E. Milner Department of Immunology, The University of Birmingham Medical School, Birmingham, England

Katherine A. Muirhead Zynaxis Inc., Malvern, Pennsylvania

Pierre Nabet Centre Hospitalier Régional Universitaire de Nancy, Nancy, France

Gerhard Nebe-von Caron Immunology Department, Colworth Laboratory, Unilever Research, Sharnbrook, Bedford, England

Steve Oh Scientific Laboratory Services, Pall Filtration Pte. Ltd., Singapore

Sylvain Paillasson Laboratoire de Neurobiologie du Développement, Ecole Pratique des Hautes Etudes, Institut Albert Bonniot, Grenoble, France

Bernhard Palsson Aastrom Biosciences, Inc., Ann Arbor, Michigan

[*]*Present affiliation*: Institut für Biotechnologie, Eidgenossische Technische Hochschule, Zurich, Switzerland.

Danilo Porro Dipartimento di Fisiologia e Biochimica Generali, Sezione Biochimica Comparata, Università degli Studi di Milano, Milan, Italy

Xavier Ronot Laboratoire de Neurobiologie du Développement, Ecole Pratique des Hautes Etudes, Institut Albert Bonniot, Grenoble, France

Sue A. Rummel Aastrom Biosciences, Inc., Ann Arbor, Michigan

Jatin D. Shah School of Chemical Engineering, The University of Birmingham, Birmingham, England

N. H. Simpson School of Chemical Engineering, The University of Birmingham, Birmingham, England

Kirsten Skarstad Department of Biophysics, Institute for Cancer Research, Oslo, Norway

Harald B. Steen Department of Biophysics, Institute for Cancer Research, Oslo, Norway

Hong Wang Department of Immunology, The University of Birmingham Medical School, Birmingham, England

James C. Weaver Harvard-MIT Division of Health Sciences and Technology, Massachusetts Institute of Technology and One Cell Systems, Inc., Cambridge, Massachusetts

Jonathan P. Welsh School of Chemical Engineering, The University of Birmingham, Birmingham, England

Sture Wold Department of Biophysics, Institute for Cancer Research, Oslo, Norway

1

Monitoring the Proliferative Capacity of Cultured Animal Cells by Cell Cycle Analysis

V. Leelavatcharamas, A. Nicholas Emery, and Mohamed Al-Rubeai

The University of Birmingham,
Birmingham, England

I. AN INTRODUCTION TO CELL CYCLE DYNAMICS IN CULTURED CELLS

The advancement of the technology for large-scale cultivation of animal cells has relied on the implementation of strategies and techniques that can be grouped into four categories: maximization of viable cell concentration, improvement of medium formulation, implementation of high-performance reactor configurations, and maximization of specific production rate. However, the development of optimal-processing schemes has always been hindered by a lack of the equipment necessary for adequate state identification in the bioreaction processes. Most of the monitoring systems available are biochemically based, slow, off-line, and yield only global values for the reaction system. Even on-line analytical systems common in bacterial and fungal fermentations, such as the measurement of oxygen uptake rates and carbon dioxide production rates by off-gas analysis, are problematic in animal cell culture owing to the very low rates to be measured and consequent poor sensitivity.

Efficient and effective process monitoring of the identified major parameters that characterize growth is the first step toward achieving adequate and sensitive process design and control. The major constituent of growth is the cell cycle, the analysis of which may provide a reliable index for the prediction of growth potential and changes in cell number with time during the cultivation period. As the cell population at anytime during cultivation is a mixture of dividing cells and dying cells, for one to predict the cell number long before it can be assessed by cell counting it is necessary to separate the component of cell growth (sometimes referred to as the growth fraction) from that of cell death. In conditions of little or no cell loss, such as are encountered during the exponential growth phase of batch culture, the analysis of the cell cycle, specifically the fraction of S cells, can furnish reliable information on the proliferative dynamics of cell culture [1]. To indicate the ability of a feeding control strategy based on cell cycle analysis to meet the demands of cells for optimal growth in perfusion culture, Miltenburger's group successfully used analysis of the percentage of S-phase cells as an indicator for regulation of the medium flow rate to hybridoma cells [2].

Such analyses of the cell cycle and, indeed, of many other structural and physiological characteristics of individual cells in a population can be efficiently, rapidly, and accurately achieved only by flow cytometry (FC). Flow cytometry can provide the detailed and thorough analyses needed for the employment of strict control of the heterogeneous cell populations normally seen in cell culture.

There is now a widespread and rapidly growing interest in the use of FC for the monitoring of large-scale animal cell bioprocesses. In this chapter we present a number of examples of its use, even when the measurement is of only a single parameter (DNA), to obtain useful insights into the proliferation kinetics of Chinese hamster ovary (CHO) cells. This parameter, which can be easily and rapidly analyzed, can singularly indicate the state of the culture and characterize its population dynamics; hence, it is likely to make a significant contribution to the development of sophisticated population balance models suitable for control with a high degree of accuracy.

The measurement of DNA was one of the first, and still is, one of the most widespread applications of FC, examples of which include the characterization of aneuploid malignant cells, cytotoxic drug effects on cell growth, the detection of polyploid species, and the separation of sperm containing the X chromosome from those containing the Y chromosome. The DNA measurements are achieved by staining with one of a variety of fluorochromes, such as propidium iodide (PI), ethidium bromide (EB), Hoechst 33258, plicamycin (mithramycin), and at least another 15 other dyes. The interaction between these dyes and the DNA molecules depends on not only the concentration of the free dye, but also on the concentrations of electrolytes and on temperature [3].

The *cell cycle* is defined as the interval between completion of mitosis in the parent cell and completion of the next mitosis in one or both daughter cells [4]. It can be divided into four phases. The G_1 phase is the period of a young cell, lasting from the completion of mitotic division until the start of the replication of its DNA. The S phase is the period of DNA synthesis, during which the genome is duplicated. The G_2 phase is the mature stage of the cell, lasting from the end of the genome duplication until the onset of the mitotic prophase. The M phase is the short period of mitosis, during which extensive structural changes appear, and at the end of which the division of the cytoplasm is initiated. For most cells, the durations of the S, G_2, and M phases are relatively constant lasting for 6–9, 2–5, and less than 1.0 h, respectively. G_1 is the most variable phase of the cell cycle: it can last 30 h or more in some cell lines, but can be lacking entirely in others (e.g., in one line of Chinese hamster lung cells [5]). Most cell populations consist of a mixture of dividing and nondividing fractions. The cell typically enters the nondividing state (or the G_0; [6]) from a point immediately after mitosis, but reenters the dividing state (or the cell cycle) in mid-G_1 [7]. Pardee et al. [8] divided the interval between G_0 and S phases into four subsections (Fig. 1a): a competent state (C), which can last for about 6 h; a point V, which is similar to that of cycling cells that have just completed mitosis; a control point R (the restriction point or start point [9]) beyond which growth factors are not required for progression through the cell cycle; and finally, a period of organization of the machinery for DNA synthesis (start of the S phase). The cell, after completing one cycle and reaching the next G_1, can then either proceed through another cycle or can enter the G_0 and stay there until it is stimulated. Not all cells can enter the G_0 state. Some transformed cells (and hybridomas) are examples of cycling cells for which their cell cycles lack the control point R, at which the entry into the S phase is determined (Fig. 1b). This situation is clear in many continuous cell lines in culture when depletion of nutrients results in reducing DNA and protein synthesis, thereby inducing cell death. Attempts to separate cell growth from cell death for the purpose of maintaining nondividing cells for high product productivity have been unsuccessful. These cells cannot be kept quiescent and viable over an extended period, and the increase in the G_1 cells seen in conditions of depleted nutrient is more likely to be due to a lack of precursors and energy sources necessary for protein and DNA synthesis. Linardos et al. [10] have studied the relationship of cell death to the cell cycle position and developed a model describing the steady-state growth and death rates of hybridoma cells. Their model was based on the transition probability model proposed by Smith and Martin [11] to explain the observed variability in the cell cycle. Their main observation, that the death rate in hybridoma cultures is proportional to the fraction of cells "arrested" in the G_1 phase, is in agreement with our own repeated observations in batch and continuous cultures [12]. Their use of the term *arrest* was understood to be a description of the fact

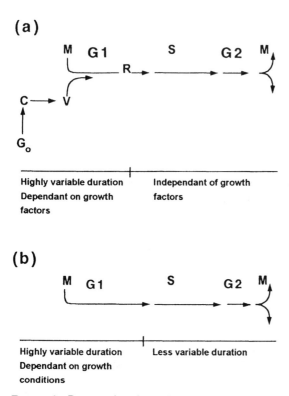

FIGURE 1 Progression through the cell cycle of normal cells (a) as proposed by Pardee et al. [8] and (b) of transformed or cancer cells for which growth factors are not critical for cell growth.

that a high percentage of G_1 phase cells were obtained at suboptimal culture conditions. The lack of nutrients results in decreased biosynthetic activity, thereby reducing cell progression. In such cultures, the only evidence of arrest is when cells are observed entirely in G_1 after nutrient depletion, or death is restricted to G_1 cells, but neither situation has ever been observed in hybridoma culture.

A typical pattern for the cell cycle distribution during simple batch culture of animal cells shows a substantial decrease in the S and G_2 fractions during the stationary and decline phases of culture. However, the increasing presence of dead cells and cell debris during these late culture phases complicates the analysis of the cell cycle distribution. To be able to use FC effectively for monitoring growth and productivity, it is first important to discriminate between viable and nonviable cells. Treatment of nonfixed cells with DNase to eliminate DNA from nonviable cells is an efficient method for the analysis of the proliferative capacity of such a heterogeneous culture [13,14].

II. SOME PROTOCOLS FOR CELL CYCLE ANALYSIS EXEMPLIFIED USING A CULTURE OF RECOMBINANT CHINESE HAMSTER OVARY CELLS

A. Cell Culture Protocol

The recombinant CHO 320 cell line used in the work described in the following, obtained from the Wellcome Foundation Ltd. (Beckenham, UK), expressed human interferon gamma, which was coamplified with dihydrofolate reductase by methotrexate selection.

The inoculum, which was taken from the late exponential phase of a 5% fetal calf serum (FCS)-supplemented culture, was spun down and resuspended in 255 ml RPMI 1640 with 1 μM methotrexate medium. In a typical experiment, in which the specific growth rates of a set of cultures were investigated by using a range of FCS concentrations, 50 ml of this suspension culture was inoculated into each batch of 1.0, 2.5, 5.0, 7.5, or 10.0% FCS culture to give the initial cell number of 2×10^5/ml in 200 ml working volume. All the cultures, in stirred bottles, were incubated at 37°C in an incubator and magnetically stirred at an agitation rate of 150 rpm. Samples were taken every 6 h in the first 2 days and every 12 h for a further 2 days after incubation. After this, samples were taken every 24 h until the end of the batch. Samples were cell-counted and centrifuged at 1000 rpm for 5 min. The cells were fixed at 10^6/ml with cold 70% ethanol and kept in the freezer for flow cytometry analysis.

B. Cell Synchronization

In another experiment, a thymidine block was used to achieve synchronization of the cell cycle. Stock thymidine, 100 mM, was added to exponentially grown CHO cells to give a final concentration of 3 mM. The culture was then incubated for 18–24 h before releasing from the block by spinning the cells down and washing them in warm, fresh medium once before resuspension in fresh medium.

C. DNase Treatment for the Elimination of Dead Cells from FC Analysis

About 10^6 cells were spun down at 1000 rpm for 5 min and resuspended in 1 ml of DNase solution. This solution was prepared according to Frankfurt [11], by dissolving 100 mg DNase (Sigma, Poole, UK) in 200 ml SMT (2.43 g Tris, 85.6

g sucrose, and 1.01 g $MgCl_2$ in 1 L distilled water, pH 6.5). The cells were incubated at 37°C for 15 min and then placed on ice for 5–10 min before they were rinsed with phosphate-buffered saline (PBS) and fixed in cold 70% ethanol.

D. DNA Staining for Flow Cytometric Analysis

Fixed cells were double washed with PBS and the cell pellet resuspended in 500 µl of 1 mg/ml ribonuclease solution (Sigma, Poole, UK) and incubated at 37°C for 20 min. The cells were then double washed with PBS, resuspended in 1 ml of 50 µl/ml of PI solution in PBS and incubated at room temperature for 15 min. The cells were washed with PBS before they were analyzed.

E. Flow Cytometric Analysis

Our flow cytometric analysis was carried out with a Coulter EPICs Elite analyzer using an air-cooled argon laser, operating at 15-mW light power, with excitation at 488 nm. A 488-nm long-pass filter was used to block the scattered laser light from reaching the fluorescence detectors. Propidium iodide fluorescence emission at 550–700 nm, with a maximum at 620 nm, was reflected to a photomultiplier tube by a long-pass 600-nm dichroic mirror and collected using a 635-nm band-pass filter. The flow cytometer was calibrated by analyzing a sample of fluorescent beads for which the 50% coefficient of variance was adjusted to be less than 2%.

Analyses of the cell cycle were based on selective gating by means of dual multiparameter analysis of the peak height and integral PI fluorescence of 10,000 cells. Figure 2 shows a typical DNA distribution in a CHO cell population during the exponential phase that was analyzed by constructing a gate to eliminate the cellular debris and aggregates. The integral fluorescence of the cells inside the selected gate was analyzed by the MULTICYCLE (Phoenix Flow Systems) computer software to obtain the percentages of cells in the G_1, S, and G_2/M compartments of the cell cycle.

III. THE RESULTS OF CELL CYCLE ANALYSIS

A. Cultures Growing at Different Growth Rates

The growth curves of CHO cells grown in the presence of different concentrations of FCS are presented in Figure 3. Both cell growth rates and cell yields increased with increasing percentage of FCS, showing both stoichiometric and kinetic limitation by substrate.

The percentage of cells in the G_1, S, and G_2/M phases for each percentage FCS concentration are illustrated in Figure 4. A regular trend in the proportions of cells in the G_1, S, and G_2/M phases can be seen, which actually shows a

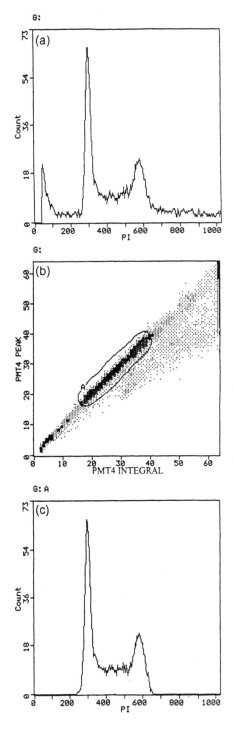

FIGURE 2 Dual parameter analysis of PI peak vs. integral to eliminate cellular debris and cell aggregates. (a) Analysis of the PI integral fluorescence of CHO cell sample from exponential phase; (b) selective gating of single cells (bitmap A) for subsequent analysis; (c) analysis of the PI integral of the selected cells inside the bitmap A.

7

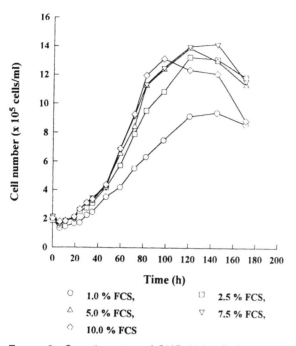

FIGURE 3 Growth curves of CHO 320 cells in media supplemented with different concentrations of fetal calf serum.

partial synchronization pattern. It is seen that a large proportion (70%) of cells in the inoculum were in G_1 because they were taken from the very late exponential phase of the inoculum culture when the culture conditions can be suboptimal owing to nutrient depletion and accumulation of cytotoxic products. After inoculation, the first sign of cell activity is a concerted move into S phase (50–60%) and subsequently, this is followed by a still largely synchronous progression, through G_2 (28–45%) and on to the first cell division. Examination of the instantaneous growth rates determined on cell number shows that this is where the sudden apparent increases in specific growth rates are seen, because most of the cells then divide at the same time. After this first cycle, the G_1 and S phase in the beginning of the second cycle can still be seen (the second peak at each phase). After that the cells become less synchronous until the growth becomes limited. These phenomena are reproduced, at least qualitatively, for all the FCS concentrations used.

The time interval between the first and second increases of the S-phase curves seen in Figure 4 (20 h) was taken as the total cell cycle duration. This value can be compared with that determined from a plot of $\ln N$ versus t in the

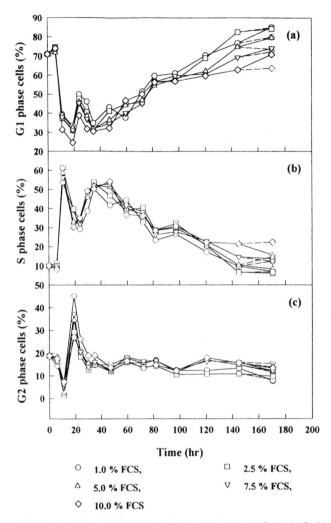

FIGURE 4 The percentage of CHO cells in the G_1 (a), S (b), and G_2 (c) phases during the batch growth in media supplemented with different concentration of fetal calf serum: solid line, with DNase treatment; broken line, without DNase treatment.

same culture for which a μ_{max} value of 0.031 h^{-1} can be calculated, giving a doubling time of 22 h. The difference probably reflects the variation in the rate of progression through the cell cycle at different times of the batch culture.

At the end of the first cycle (about 24 h in Figs. 3 and 4), the effects of serum concentration on cell growth were clearly shown. The higher serum concentration gave the higher growth rate; consequently, more cells in the high serum concentration progressed through the cell cycle and, thus, resulted in the higher percentage of G_2 seen here. The difference in specific growth rate can arise from the difference in the cell cycle length; the shorter the cycle time the higher the growth rate. This type of variation is dependent on the physiological conditions (hormones, growth factors, nutrients, or other). In addition, in a population the individual cells have variable cell cycle times [4]. Among the cell cycle phases, the length of the G_1 phase is the more variable, whereas those of the S, G_2, and M phases are relatively constant. In these cases, the lower serum concentration presumably lead to a longer G_1 phase. However, these results do not exclude the possibility that there are probably variable-rate processes throughout the cell cycle.

Clearly these results, even though for the single parameter, give substantially more indication of the state of the culture than is available from simply measuring cell numbers and metabolite concentrations. They have shown (a) an inoculum state, (b) the changing proliferative state of the culture, and (c) that growth rates based on cell numbers may well differ when calculated over short times, from those based on cell mass or volume.

B. Monitoring the Response to Chemically Induced Synchrony

Chemically induced synchrony by either single- or double-block thymidine treatment was not successful and resulted in only a similar "partial synchronization" to that described in the foregoing Sec. III.A. Treatment of cells with 3 m*M* thymidine for 18 h resulted, 3 h after the release from the block, in the accumulation of 76% of cells in the S phase (Fig. 5). After 7 h, the cell number started to increase, as did the fraction of cells in the G_1 phase. The peak of the G_1 phase, about 50%, was found at 10 h after release, and the cell number had then risen to 5.8×10^5 from 4×10^5/ml. Cell cycle duration was slightly shorter than that found for batch cultures.

One obvious trend seen in Figure 4 is that, on increasing the incubation time, the proportion of S phase cells was reduced, whereas the proportion of cells in the G_1 phase was increased. The increase in the G_1 phase cells and the decline in the S phase cells corresponded with the decrease in the specific growth rate. This behavior was also found in hybridoma cells [15]. The proportion of G_2 phase cells, however, was constant after a temporary increase at the beginning. These results, however, suggest that a good degree of permanent synchronization could be achieved by routine subculture, which takes into account the subculture interval and medium condition. Permanent synchronization would be advantageous in some cell lines for the maximization of physiological activity and product productivity.

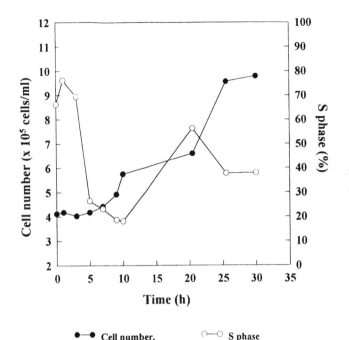

FIGURE 5 Cell number and percentage of S phase cells of CHO cells after release from a single thymidine block.

Here FC has provided confirmation of what would otherwise be an extremely difficult experimental observation. Indeed, "degree of synchrony" is a concept virtually impossible of description without the application of FC.

C. The Effects of DNase on Flow Cytomentric Cell Cycle Analysis

We have previously suggested [12] that the dead cells (especially necrotic cells), in which the DNA distribution could not be distinguished from that for the live cells, could cause difficulties in interpretation of the DNA histograms. Serious underestimation of the proportions of S and G_2 cells could result owing to the reduced PI fluorescence, thereby introducing a bias in the description of the growth and production kinetics. To demonstrate this, and also the effect of an experimental protocol designed to remove this bias, at the end of each batch two samples of cells were treated, one with, and the other without, the enzyme DNase for 30 min at 37°C before fixing with ethanol. The results are shown in Figure 6, in which the differences between samples with and without DNase

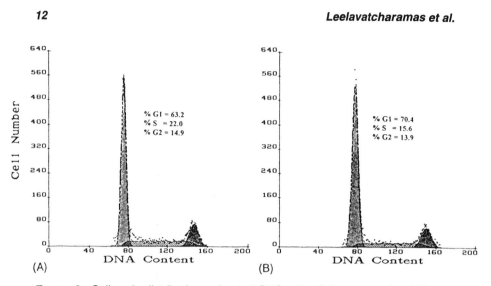

(A)   (B)

FIGURE 6 Cell cycle distributions of gated CHO cells. Cells were stained (A) with PI without DNase or (B) with DNase.

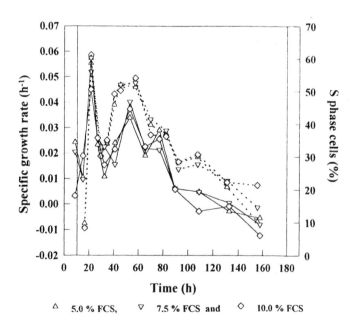

FIGURE 7 A comparison of the proportion of S phase cells (dotted lines) and the specific growth rate (solid lines) measured 10.5 later—shown by moving the time axis for the latter 10.5 h out of phase.

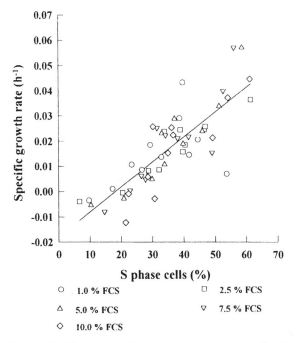

FIGURE 8 The correlation between percentage S phase cells and the specific growth rate measured 10.5 h later for all CHO batch cultures.

treatment are observed. Without treatment with DNase, the proportion of S cells in the death phase would be overestimated. The removal of S and G_2 cells by the DNase treatment indicates that cells die at any phase of the cell cycle and do not need to be "arrested" in the G_1 before they die. Fortunately, such problems arise only in the decline phase, when dead cells accumulate because of nutrient depletion and other stresses.

D. Use of Cell Cycle Analysis for the Prediction of Future Growth Rates

Measurement of the proportion of S cells has been used to enable prediction of changes in hybridoma cell number following perturbation of the culture condition [15]. From several results with CHO cultures, the proportion of cells seen at any time in the S phase can give a good indication of future specific growth rates. If we move a typical curve of instant specific growth rate versus time 10.5 h out of phase with the curve for the proportion of S phase cells (Fig. 7), we see a reasonably close comparison. Given all our data, the correlation (Fig. 8) for all the FCS concentrations, can be used as a general indication of the future status

of the culture. The variability is not surprising, given (a) the short-term variations in μ seen, owing to the synchrony referred to earlier, and (b) the accepted
high variability of microscopic cell counts based on Trypan blue staining. These
results support the claims of Kloppinger et al. [16] and Al-Rubeai et al. [17] that
FC-based cell cycle analysis is a feasible method for the provision of advanced
and reliable information on proliferation dynamics in cell culture. Later chapters
will demonstrate just how much more information on culture state can be
derived simultaneously and rapidly using FC, thereby providing a comprehen-
sive information base for the application of sophisticated control and decision
making. It may be reasonably expected that such systems, particularly those
based on intelligent, "self-learning" principles, will be widely available in
coming years.

IV. CONCLUSIONS

The FC-based analysis can show that normal conditions of preparation and use
of an inoculum can induce a partial synchronization growth pattern at the beginning of a batch culture of cells. Such effects have been reproduced for several
different specific growth rates in the early to mid-exponential phases of batch
culture of CHO cells.

 Dead cells and cell debris, especially in the death phase of growth, can
cause a wrong estimation of DNA distributions during the batch culture. However, this can be avoided by treating cells with DNase before cell fixing.

 The analysis of cell cycle by FC can be used as an indicator in predicting
the future status of the culture. The longer the incubation period, the lower the
proportion of cells in the S phase and the higher the proportion of G_1 cells. A
useful linear relation between the proportion of S phase cells found using FC and
the specific growth rate later can be found, demonstrating the advantages of FC
in giving early indication of a future process state.

 All these examples illustrate how even a simple classic application of FC
can generate information and experimental opportunities not otherwise available.

REFERENCES

1. M. Al-Rubeai and A. N. Emery, Flow cytometry in animal cell culture, *Biotechnology 11*:572 (1993).
2. M. Al-Rubeai, M. Kloppinger, G. Fertig, A. N. Emery, and H. G. Miltenburger,
   Monitoring of biosynthetic and metabolic activity in animal cell culture using flow
   cytometric methods, *Animal Cell Technology* (R. E. Spier, J. B. Griffiths, and C.
   MacDonald, eds.), Butterworth-Heinemann, Oxford, 1992, p. 301.

3. J. Kapuscinski and Z. Darzynkiewicz, Special properties of fluorochromes used in flow cytometry. *Methods in Cell Biology*, Vol. 33, *Flow Cytometry* (Z. Darzynkiewicz and H. A. Crissman, eds.), Academic Press, San Diego, 1990, pp. 655–669.
4. R. Baserga, *The Biology of Cell Reproduction,* Harvard University Press, Cambridge, 1985.
5. E. Robbins and M. D. Scharff, The absence of a detectable G_1 phase in a cultured strain of Chinese hamster lung cells, *J. Cell Biol. 34*:684 (1967).
6. L. G. Lajtha, On the concept of the cell cycle, *J. Cell Comp. Physiol. 62:*143 (1963).
7. R. Baserga and E. Surmacz, Oncogenes, cell cycle genes and the control of cell proliferation, *Biotechnology 5:*355 (1987).
8. A. B. Pardee, D. L. Coppock, and H. C. Yang, Regulation of cell proliferation at onset of DNA synthesis, *J. Cell Sci. Suppl. 4:*171 (1986).
9. A. W. Murry, The cell cycle as a cdc2 cycle, *Nature 342:*14 (1989).
10. T. I. Linaedos, N. Kalogerakis, and L. A. Behie, Cell cycle model for growth rate and death rate in continuous suspension cultures, *Biotechnol. Bioeng. 40:*359 (1992).
11. J. A. Smith and L. Martin, Do cells cycle? *Proc. Natl. Acad. Sci. USA 70:*1263 (1973).
12. M. Al-Rubeai, A. N. Emery, S. Chalder, and D. C. Jan, Specific monoclonal antibody productivity and the cell cycle—comparisons of batch, continuous and perfusion cultures, *Cytotechnology 9:*85 (1992).
13. M. Al-Rubeai and A. N. Emery, Process application of flow cytometry to the analysis of growth and physiological response in animal cells, ECB6: Proceedings of the 6th European Congress on Biotechnology (L. Alberghina, L. Frontali, and P. Sensi, eds.), Elsevier Science, Amsterdam, 1994, pp. 587–591.
14. O. S. Frankfurt, Flow cytometric measurement of cell viability using DNase exclusion, *Methods in Cell Biology*, Vol. 33, *Flow Cytometry* (Z. Darzynkiewicz and H. A. Crissman, eds.), Academic Press, San Diego, 1990, pp. 13–18.
15. M. Al-Rubeai, S. Chalder, R. Bird, and A. N. Emery, Cell cycle, cell size and mitochondrial activity of hybridoma cells during batch cultivation, *Cytotechnology 7*:179 (1991).
16. M. Kloppinger, G. Fertig, E. Fraune, and H. G. Miltenburger, Flow cytometric process monitoring of hybridoma cells in batch and perfusion culture, *Advances in Animal Cell Biology and Technology for Bioprocesses* (R. E. Spier, J. B. Griffiths, J. Stephenne, and P. J. Croony, eds.), Butterworths, London, 1989, pp. 125–128.
17. M. Al-Rubeai, A. N. Emery, and S. Chalder, Flow cytometric analysis of cultured hybridoma cells, *J. Biotechnology 19*:67 (1990).

2

Use of Flow Cytometry for Monitoring Antibody Productivity and Isolating High-Secreting Hybridoma Cells

KATHERINE L. MCKINNEY* AND GEORGES BELFORT
Rensselaer Polytechnic Institute, Troy, New York

I. INTRODUCTION

Flow cytometry (FC) has become a useful tool aiding in the analysis and optimization of in vitro cell culture processes. It has been widely used to study antibody-producing hybridoma cells, increasing knowledge about the physiological state of cells under various conditions and at various times during their cultivation. Also, FC has revealed heterogeneity within cell populations, leading to the possible isolation of cell line variants with desirable characteristics.

Researchers have attempted to correlate or show no association between various parameters measured by flow cytometry or by other techniques during cultivation of hybridoma cells. These include surface-associated antibody, total cell-associated antibody, RNA levels, total protein levels, cell size, growth rate, and level or rate of secreted antibody [1–11]. Observation of increased antibody production during particular phases of the cell cycle has also been reported [9,12–17]. Even though correlations have often been contradictory among

* *Present affiliation*: Institut für Biotechnologie, Eidgenossische Technische Hochschule, Zurich, Switzerland.

17

different researchers using different cell lines, a better understanding of what factors affect antibody production and how to use these factors to effect increased production has been gained. For example, strategies including separation of high- and low-producing cells, limiting the fraction of low-producing cells with feeding or growth strategies, arresting cells in G_1 phase with slow growth, and growing cells at high density under low-growth conditions with perfusion of medium, all have been successful. Also, subjecting cells to environmental stresses, such as changes in medium osmolarity, pH, temperature, dissolved oxygen concentration, or hydrodynamic stresses, has also increased antibody secretion rates [13,18–20]. As a result of these studies and others using FC, a greater understanding of the dependence of productivity changes on physiological state of the cells has been gained.

Flow cytometry has also been useful in identifying heterogeneity within hybridoma cell populations. Many researchers have observed a loss in antibody productivity with culture age, as well as a shift in the level of intracellular antibody [12, 21–28]. Often a subpopulation of cells emerges with lower intracellular antibody concentration, creating a bimodal intracellular antibody distribution. This decrease in productivity may be due to chromosomal mutation or loss, to depletion of critical serum factors during adaptation to serum-free medium, or to decrease in the rate of transcription of heavy and light chain genes or stability of mRNA. Selection of high-producing cells or elimination of low-producing cells by fluorescence-activated cell sorting (FACS), therefore, is an important optimization concern, especially for large-scale antibody production.

In this chapter, the pertinent flow cytometric staining procedures will first be described, followed by the use of flow cytometry for monitoring antibody production under different experimental conditions. Finally, isolation of high-secreting hybridoma cells will be discussed.

II. FLOW CYTOMETRIC STAINING PROCEDURES

Murine hybridoma cells DB968 (ATCC HB124, Rockville, MD) or CVC.7 (ATCC TIB138) producing IgG2a against bovine insulin and bovine brain clathrin, respectively, were used in the following studies. Both cell lines resulted from fusions with Sp2/0-Ag14 cells. Cells were cultured in RPMI-1640 with 10% fetal bovine serum (FBS) and supplements or hybridoma serum-free medium (HSFM) (Gibco, Grand Island, NY). Supernatant antibody concentrations were determined using an enzyme-linked immunosorbent assay (ELISA) or protein A perfusion chromatography (PerSeptive Biosystems, Cambridge, MA) [29].

Analysis of surface and intracellular antibody immunofluorescence and DNA distributions were performed on EPICS C, EPICS 752 (Coulter Electronics, Hialeah, FL), or FACSCAN (Becton Dickinson Immunological Systems,

San Jose, CA) flow cytometers. Excitation of cell samples at 488 nm from argon lasers was employed. The cytometers were calibrated with 10-μm fluorescent beads (Coulter Electronics, Hialeah, FL), with a coefficient of variance less than 2%, before running samples each day.

A. Surface-Associated Antibody

Antibody molecules associated with the surface of the cells were stained with fluorescein isothiocyanate (FITC) -conjugated goat antimouse (GAM) IgG1 for the negative control sample and FITC GAM IgG2 for the positive sample. A FITC-conjugated rabbit antimouse (RAM) IgG2b and FITC RAM IgG2a (ICN, Lisle, IL) were used in other experiments as the negative and positive stains, respectively. Also, a FITC RAM IgG1 (ICN, Lisle, IL) was used for control samples during staining of total cell-associated IgG as needed.

In the surface-staining procedure [30], the cells remained viable and were kept sterile. The cells were initially washed twice with cold phophate-buffered saline (PBS) containing 2% fetal bovine serum and then incubated on ice with the appropriate antibodies for approximately 30 min. The cells were washed again in the cold buffer and analyzed immediately on the flow cytometer. The negative control sample generates a peak, accounting for nonspecific binding of the antibodies and cellular autofluorescence. Negative and positive samples were run separately, and 10,000 cells from each sample were typically analyzed. The percentage positive values are reported for surface-stained cells.

B. Total Cell-Associated Antibody

A dual-staining procedure [21] allowed simultaneous analysis of intracellular antibody content and DNA distribution. The cells were fixed with ethanol and stained with the appropriate antibodies for negative and positive samples. Propidium iodide (PI; Sigma Chemical Co., St. Louis, MO) was used to stain the DNA at a concentration of 20 μg/ml. Mean peak fluorescence values are reported for intracellular staining.

The HB124 cells grown in HSFM were stained on the surface and inside with the same fluorescent-labeled antibodies to determine how much surface antibody fluorescence contributes to the total cell-associated antibody fluorescence. The FITC RAM IgG1 (negative sample) and FITC RAM IgG2a (positive sample) were used in the aforementioned protocols for both surface and intracellular staining. All samples were run at cytometer settings normally used for intracellular samples. The percentage of surface antibody fluorescence that contributes to total antibody fluorescence was calculated to be approximately 3%. Therefore, large changes in surface antibody fluorescence may be observed with small changes in total cell-associated antibody fluorescence. Total cell-associated and intracellular antibody fluorescence, therefore, can be used interchangeably.

C. Dual-Color Surface and Intracellular Antibody

To determine whether a correlation exists between surface and intracellular antibody, cells (both HB124 and TIB138 grown in HSFM) were stained with an FITC-labeled antibody (green) on the surface and a phycoerythrin (PE)-labeled antibody (red) inside the cell. For each sample, 5×10^5 cells were washed with PBS buffer and resuspended in dilutions of RAM IgG2b (negative sample) or IgG2a (positive sample).

Samples were incubated on ice for 30 min, washed two times, and resuspended in 2% paraformaldehyde in PBS. Cells were incubated for 15 min on ice with frequent mixing. The paraformaldehyde cross-links antibody on the cell surface before the permeabilization step with ethanol. Cells were washed once and resuspended in 70% ethanol and incubated on ice for 10 min. Samples were then washed twice and resuspended in dilutions of streptavidin PE (Vector, Burlingame, CA) (negative sample), or streptavidin PE plus GAM IgG biotin F(ab')$_2$ (Jackson ImmunoResearch, West Grove, PA) (positive sample) for 30 min on ice. Cells were washed twice before analysis by FC.

III. MONITORING ANTIBODY PRODUCTIVITY USING FLOW CYTOMETRY

A. Analysis of Hybridoma Cultures with Different Inoculum Densities

Surface antibody fluorescence has been studied by many researchers who used different cell lines, which has resulted in contradictory data. Meilhoc et al. [2] observed two populations relative to surface antibody fluorescence; however, the specific antibody secretion rate did not correlate well with surface fluorescence or total cell-associated IgG. In contrast, Sen et al. [5] found a linear correlation between mean surface fluorescence and the specific antibody production rate. Interestingly, Ozturk and Palsson [31] found little difference in antibody production rate for cells grown in batch culture at varying initial cell densities (but only from 10^3 to 10^5 cells/per milliliter).

In the following experiments, flow cytometry was used to monitor surface antibody fluorescence and used as an indicator of antibody productivity. Cells were seeded at densities of 1×10^4, 5×10^4, 1×10^5, 5×10^5, and 1×10^6 viable cells per milliliter in T-flasks. The inoculation procedure involved centrifuging the cells and resuspending them in the appropriate medium, as the study was conducted in both conditioned and unconditioned media. The conditioned medium was prepared by seeding 250-ml spinner flasks with 5×10^4 viable cells per milliliter and growing them for 2 days. The media samples were aseptically collected and centrifuged for use in the experiment. The cultures were incubated overnight, stained with FITC RAM IgG2b and IgG2a, and analyzed for surface antibody content on the flow cytometer. Samples were taken for cell count and analysis of medium antibody concentration.

Surface antibody histograms for cultures grown at different initial cell densities in unconditioned or conditioned medium showed a decrease in the surface fluorescence, with an increase in the initial cell density (Fig. 1). This is also accompanied by a decrease in the specific IgG secretion rate (Table 1). In conditioned medium, the specific secretion rate is much higher at low initial cell concentrations, when compared with unconditioned medium, and decreases to very low levels at high cell density. It is expected that the conditioned medium is rich in important growth factors and nutrients, leading to high initial secretion rates, whereas depletion of nutrients at high cell density causes the abrupt drop in antibody secretion.

Therefore, density of cells in the culture flask influenced cell surface fluorescence and antibody productivity. Very low cell densities showed the highest surface fluorescence, which decreased with increasing cell density. As expected, increased antibody yield was observed for cultures with higher initial cell densities. Culturing cells under conditions at which their secretion rates are highest may not correspond to the highest medium antibody yield obtainable; however, consideration of environments in which the cells are likely to overproduce antibody will be important in designing new cell culture methods.

B. Analysis of Batch Hybridoma Cultures

To extend some of the foregoing observations to longer-term experiments, batch culture of hybridoma cells was investigated. Hybridoma cell lines HB124 and TIB138 were grown in HSFM (Hybridoma Serum-Free Medium, GibcoBRL, Grand Island, NY) over a 7-day period (day 0–7). Flow cytometric analyses of surface antibody fluorescence and intracellular antibody fluorescence were carried out on days 1 through 5. Samples for determination of cell count, viability, intracellular antibody concentration, and supernatant antibody concentration were taken daily. Intracellular antibody concentration was determined following the procedure of Meilhoc et al. [2]. Cells were stained for surface and intracellular IgG as outlined. The FITC-conjugated RAM IgG2b, IgG1, and IgG2a (ICN, Lisle, IL) were used to target the produced antibodies.

Figure 2 shows that surface antibody content and specific antibody secretion rate decline simultaneously in batch culture. The TIB138 cultures displayed a bimodal distribution for intracellular antibody concentration, whereas the HB 124 cultures did not (Fig. 3). For the TIB138 line, the percentage of cells in each peak remained constant throughout the cultivation period. In Figure 4a, intracellular antibody concentration for HB124 is plotted with intracellular log mean peak fluorescence versus time. Figures 4b and 4c show the intracellular antibody concentrations and the low and high log mean peak fluorescence values for the TIB138 cells. Log mean peak values could be used to determine how

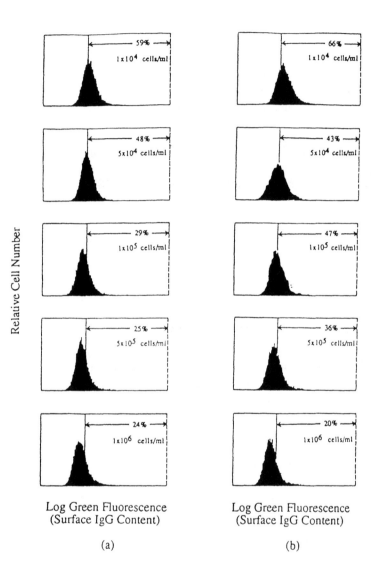

Log Green Fluorescence
(Surface IgG Content)

Log Green Fluorescence
(Surface IgG Content)

(a)   (b)

FIGURE 1 Variation of cell density in batch culture: Fluorescent surface antibody histograms for HB124 hybridoma cultures grown at different initial cell densities in (a) unconditioned and (b) conditioned media. Cells were analyzed 1 day after seeding flasks. Placement of the cursor represents the point at which cells are considered to have positive fluorescence, as the negative peaks are not shown.

TABLE 1 Fluorescent Surface Antibody Content and Antibody Production for Batch HB124 Hybridoma Cultures Grown at Various Initial Cell Densities

	Seeding density (cells/ml)				
	1×10^4	5×10^4	1×10^5	5×10^5	1×10^6
Unconditioned media					
% positive surface fluorescence[a]	59	48	29	25	24
Specific IgG secretion rate (μg/day) per 10^6 cells	90	63	83	23	14
Medium antibody concentration (μg/ml)[b]	1.7	5.7	10.6	19.8	21.4
Conditioned media					
% positive surface fluorescence[a]	66	43	47	36	20
Specific IgG secretion rate (μg/day) per 10^6 cells	590	100	40	10	6
Medium antibody yield (μg/ml)[b,c]	9.8	9.0	6.2	6.6	7.8

[a]Approximately 10,000 cells were analyzed for each sample 1 day after seeding flasks.
[b]Error associated with ELISA assay may be as high as 10%.
[c]Values are final IgG concentration minus initial conditioned medium IgG concentration.

much antibody is contained, on average, in cells in the low peak relative to those in the high peak. By knowing the fraction of cells in each peak and the average intracellular concentration, values for high- and low-peak concentrations could be determined. The subsequent information was used to determine average secretion rates for cells from each peak [11]. Low-peak cells secreted antibody at a rate approximately six times less, on average, than those in the high peak. In any event, it has become apparent from this study that large differences in secretion rates exist between low- and high-peak cells, and that it is necessary to separate the low and high secretors to maintain high antibody productivity.

Interestingly, Leno et al. [7] showed that membrane immunofluorescence followed cytoplasmic immunofluorescence and cytoplasmic antibody concentration quite closely. As cells passed from exponential into stationary growth phases, decreases in cytoplasmic and membrane-bound antibody were observed, as well as decreases in kappa- and gamma-chain mRNA concentrations. These parameters, however, did not correlate with specific antibody secretion rate throughout the culture period. Al-Rubeai et al. [6] noted a decrease in mean fluorescence intensity for DNA, RNA, protein, and intracellular IgG during the decline phase of growth. Intracellular immunofluorescence, rather than surface immunofluorescence, correlated with specific antibody productivity. Dalili and Ollis [3] reported that cellular RNA content remained constant throughout expo-

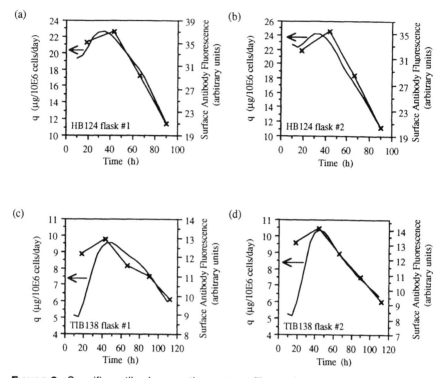

FIGURE 2 Specific antibody secretion rate profiles and surface antibody fluorescence data, reported as mean peak channel (x), for HB124 and TIB138 hybridoma cells in batch culture: (a) HB124 flask 1, (b) HB124 flask 2, (c) TIB138 flask 1, and (d) TIB138 flask 2.

nential growth phase, and then, decreased significantly as cells entered stationary phase, and that average cellular protein and antibody content decreased significantly with glutamine depletion. Thus, lower RNA content was consistent with lower levels of protein synthesis. Similarly, as a batch culture progresses, protein translation should decrease as growth rate decreases. Since rRNA, or ribosomal RNA, makes up about 80% of the RNA measured, protein translation is expected to decrease with decreases in total RNA. Bibila and Flickinger [32] reported a linear decrease in total cellular RNA content with decreasing specific growth rate during batch culture. This decrease in protein translation with decreasing growth rate should subsequently cause a decrease in expression of cell surface molecules, as well as rates of secretion of proteins, as the batch culture progresses.

(a)   HB124 flask #1

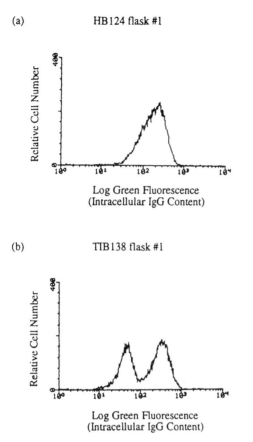

Log Green Fluorescence
(Intracellular IgG Content)

(b)   TIB138 flask #1

Log Green Fluorescence
(Intracellular IgG Content)

FIGURE 3 Intracellular antibody fluorescence for HB124 and TIB138 hybridomas cells in batch culture: (a) HB124 flask 1 and (b) TIB138 flask 1 at 19 h.

C. Effect of Solute Stress on Hybridoma Cultures

Many types of environmental stress have been shown to effect an increased specific antibody production by hybridoma cells with concurrent decrease in cell growth rate. Various studies on solute stress, using sodium chloride, ammonium chloride, and lactate, as well as pH, temperature, dissolved oxygen, and hydro-dynamic stresses, have been reported. There is potential to exploit these effects in commercial production of cell culture products; however, cellular metabolic responses to such environmental stresses are just beginning to be understood.

Increased antibody production has been observed in solute-stressed cultures. Maiorella [18] reported that inhibitory solutes could increase antibody

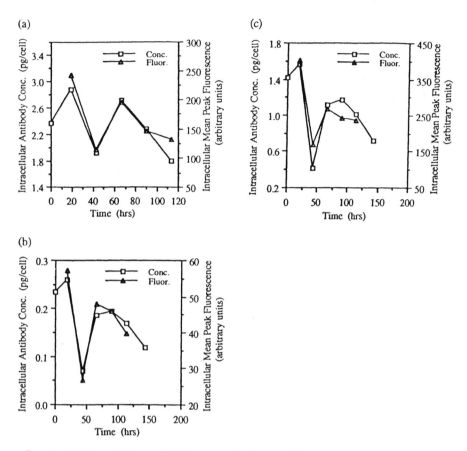

FIGURE 4 Intracellular antibody concentration and log mean peak intracellular antibody fluorescence for HB124 and TIB138 hybridoma cells in batch culture: (a) HB124 flask 1, (b) TIB138 flask 1 (low peak), and (c) TIB138 flask 1 (high peak).

titer in hybridoma cell cultures: The use of 10 mM ammonium chloride, 30 mM lactate, and addition of sodium chloride to increase medium osmolarity from 300 to 400 mOsm, resulted in antibody titers that were about twice, and cell densities that were about half, the values of those observed for the control cultures. Maiorella also reported that stressing the cells with NaCl at pH 6.8 resulted in significant increases in specific antibody secretion rate, compared with cultures maintained at normal osmolarities and pH conditions. A pH of 7.2 was optimal for growth of the particular cell line studied, whereas antibody productivity was highest at pH 6.8. Furthermore, Berg et al. [33] reported the effects of hyperosmolarity caused by addition of salts and sugars, specifically NaCl, KCl, sucrose, and xylitol, to the growth medium of hybridoma cells. Reduced growth rates and cell densities, and increased specific antibody production were observed. Ozturk and Palsson [20] also altered medium osmolarity by the addition of NaCl or PBS and sucrose to final osmolarities of 290, 338, 386, 435, and 580 mOsm. Growth rate decreased, and glucose and glutamine utilization, and lactate and ammonia production rates increased on a per cell basis with increase in osmolarity for all of the aforementioned solutes. Specific antibody productivity increased as well, with a twofold increase observed at 435 mOsm. Furthermore, Reddy et al. [34] reported that intracellular antibody content of viable cells, as determined by flow cytometry, and specific antibody secretion rate remained high during the death phase after sodium chloride addition. In control cultures, these parameters both decreased. Most recently, Oh et al. [35] reported increased immunoglobulin levels (2.3 times) by suppressing cell growth and increasing culture longevity by adapting hybridoma cells to increased osmolarity (350 mOsm) with 0.1 M sodium butyrate addition.

We have studied the effect of sodium chloride solute stress on hybridoma cells using flow cytometry. Sodium chloride was added to cultures of hybridoma cells, seeded at 4×10^4/ml, to increase the medium osmolarity. Final osmolarities were 273 (control), 336, 400, 457, and 518 mOsm. Osmolarities of 457 mOsm and higher were quite lethal for the cells, so in a second experiment, only osmolarities of 270 (control), 325, and 395 mOsm were studied. On day 3, samples were taken for analysis of surface and intracellular antibody by flow cytometry. Antibody secretion rates were calculated between day 2 and day 3.

Viable cell counts for the sodium chloride stress experiments are listed in Table 2 (up to 400 mOsm). Histograms representing surface and intracellular antibody fluorescence for the second experiment are shown in Figure 5. Table 2 shows percentage positive surface antibody fluorescence and mean peak channel intracellular antibody fluorescence values, as well as specific antibody secretion rates. For the sodium chloride solute stress, antibody secretion rate follows the percentage positive surface antibody fluorescence, which is correlated with medium osmolarity, whereas surface and intracellular fluorescence are inversely correlated. Release of stored antibody owing to leaky membranes is probably the cause of the highest secretion rate (80 μg/10^6 cells per day) noted for a stress of 395 mOsm in experiment 2.

In this study, sodium chloride solute stress increased specific antibody production (see Table 2). Because the stress also decreased the maximum cell densities obtained, the final IgG concentrations for stressed cultures were lower. Therefore, to benefit from the increased specific antibody secretion rate under

TABLE 2 Flow Cytometric Analysis and Antibody Production for HB124 Cells Stressed with Sodium Chloride in Batch Culture

	Antibody fluorescence[a]		Antibody production[b,c]		
Solute stress (day 3)	Surface (% positive)	Intracellular (mean peak channel)	Secretion rate ($\mu g/10^6$ cells/day)	Conc. ($\mu g/ml$)	Cell density (cells/ml $\times 10^4$)
NaCl (mOsm)					
Experiment 1					
273	34	130	12	18.1	37.9
336	65	104	24	18.1	27.4
400	47	112	19	16.0	9.5
Experiment 2					
270	36	104	12	23.4	39.6
325	45	106	18	23.4	35.1
395	63	92	80	18.9	11.4

[a]Flow cytometric analyses were carried out on day 3 of the batch cultures.
[b]Antibody production was analyzed on day 3 and secretion rates were calculated between day 2 and day 3 of the batch cultures.
[c]Error associated with ELISA assay may be as high as 10%.

conditions of stress, a two-step strategy could be used. Initially, the cells should be grown to the desired cell density, and then, the osmolarity of the medium increased to boost antibody secretion rates. The effects of solute stress would have to be tested at higher cell densities at which cells would be growing at slower rates. Surface antibody fluorescence followed specific antibody secretion rate, a result that has also been observed in many other experiments. It is especially interesting that surface antibody fluorescence, which is thought to follow the general level of protein translation, is increased during a period of stress (see Fig. 5) and, therefore, it was not positively growth-associated. Also, interestingly, surface and intracellular antibody fluorescence are inversely correlated. At the highest surface antibody levels, the lowest intracellular antibody levels are observed. The solute stress, therefore, may cause a decrease in growth rate, which allows significantly more energy to be available for processing and secretion of antibodies, or for release of stored antibody from inside the cell; hence, the decrease in intracellular antibody. Furthermore, the condition of the stressed cell may not permit rapid assembly of more antibodies to be processed and secreted. A similar decrease is seen as cells in batch culture enter the stationary phase of growth (see Fig. 4).

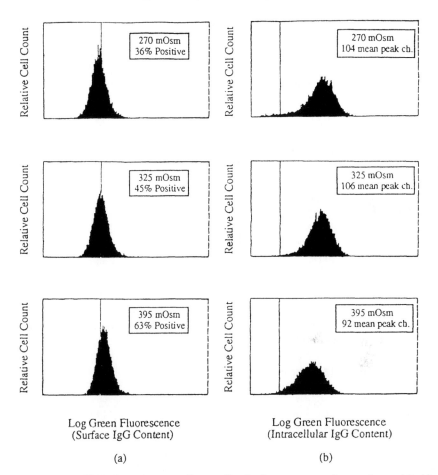

Log Green Fluorescence
(Surface IgG Content)

Log Green Fluorescence
(Intracellular IgG Content)

(a)

(b)

FIGURE 5 Histograms representing antibody fluorescence for sodium chloride solute stress (270, 325, and 395 mOsm) of HB124 cells: (a) surface antibody fluorescence and (b) intracellular antibody fluorescence. The cursor marks the point above which cells are considered to have positive fluorescence, as negative peaks are not shown.

IV. FLUORESCENCE-ACTIVATED CELL SORTING

A. Bimodal Intracellular Antibody Distribution

Fluorescence-activated cell sorting has enabled large quantities of cells with desired characteristics to be rapidly isolated. The criteria for selection of high-producing cells by flow cytometric methods have been based on the isolation of high-producing cells within gel microdroplets or of cells containing an increased amount of surface antibody. In the first method, single cells are encapsulated in an alginate or agarose matrix [36]. The product secreted by each cell is captured near the cell and can be quantified. The second technique involves staining the surface of the hybridoma cells directly with fluorescent-labeled antibodies and selecting for those cells displaying an increased amount of surface fluorescence [1, 4, 8]. This method enables rapid and efficient sorting of large quantities of high-producing cells, provided that surface fluorescence is correlated with the antibody secretion rate.

To sort viable cells, a correlating parameter that represents intracellular antibody content is necessary. Figure 6 shows the results of a dual-staining procedure for surface and intracellular antibody for both the HB124 and TIB138 cells. The HB124 cells are clustered well on the two-color plot (see Fig. 6a) showing that the parameters are not correlated under the conditions investigated. In Figure 6b (TIB138 cells), however, the two-color plot shows that cells with low intracellular IgG also have low surface IgG relative to the cells with high intracellular IgG. Therefore, surface antibody fluorescence can be used to select for the higher-producing cells, as long as substantial differences exist among intracellular antibody concentrations, signifying differences in specific antibody secretion rates within the population.

B. Sorting High-Secreting Cells by Surface Antibody Fluorescence

In addition to using FITC-conjugated anti-IgG2a antibodies, antibody on the surface of the cells was labeled using fluorescent latex beads, 0.79 μm diameter (Baxter Healthcare, Mundelein, IL) that had been labeled with RAM IgG1 and IgG2a (Zymed Laboratories, South San Francisco, CA). The cells were stained by layering the bead solution over cell suspensions contained in 24-well plates. The plates were centrifuged, forming a monolayer of cells coated with beads [30, 37], before analysis by FC.

Bitmap gates, or windows, were created that enabled sorting a percentage of the cells with the highest and lowest surface antibody content. Bitmap windows were drawn using forward-angle light scatter (FALS) representing cell size and 90° light scatter representing cell granularity, for surface-stained cells to exclude doublets, cell debris, and nonviable cells from the analysis. Sorting by surface antibody content was performed by drawing bitmap windows based on cell size and green fluorescence signals (Fig. 7). Care was taken so that only cells of the same size were sorted.

The surface antibody content of hybridoma cells stained with FITC GAM IgG1 and IgG2 was analyzed, and cells were sorted by FC. Separate flasks con-

(a)

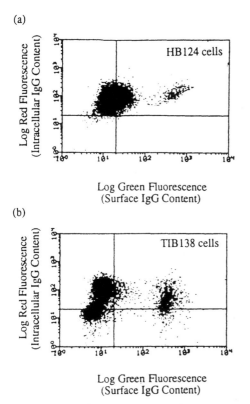

Log Green Fluorescence
(Surface IgG Content)

(b)

Log Green Fluorescence
(Surface IgG Content)

FIGURE 6 Dual stain for surface antibody fluorescence (green) and intracellular antibody fluorescence (red) for HB124 and TIB138 hybridoma cells in batch culture in serum-free medium: (a) positive red / positive green fluorescence for HB124 cells, and (b) positive red / positive green fluorescence for TIB138 cells. Cursor locations mark boundaries between cells exhibiting negative and positive fluorescence.

taining 4% of the cells with the highest and lowest surface antibody, as determined by surface fluorescence measurements, were obtained. An unsorted population of cells was also collected by running cells through the flow cytometer and excluding nonviable cells from the collected population. Care was taken to obtain the same number of cells in each flask so that all cultures would be subjected to similar conditions. Cells were cultured in 250-ml T-flasks, and samples were taken each day over a 6-day period to determine cell count, viability, and medium antibody concentration. Surface antibody content was measured with the FITC-labeled antibodies on day 4 of each batch culture. The surface

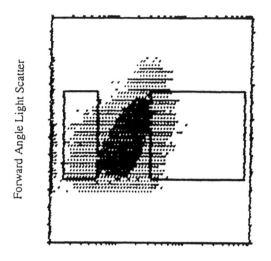

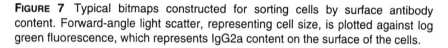

Log Green Fluorescence

FIGURE 7 Typical bitmaps constructed for sorting cells by surface antibody content. Forward-angle light scatter, representing cell size, is plotted against log green fluorescence, which represents IgG2a content on the surface of the cells.

antibody measurement obtained by staining cells with the FITC-labeled antibody did not give as strong a signal as was desired for separation of the high from low surface antibody-containing cells. Consequently, fluorescent beads were used in place of the FITC-conjugated antibody, and cells were sorted by FC. Use of the fluorescent beads as the surface antibody marker expanded the fluorescence level of the positive sample farther beyond the control, which enabled slightly better separation of the high and low producers (Table 3).

As seen in Figure 8, after 4 days in culture, cells still displayed the characteristic fluorescence by which they were sorted; however, they had begun to revert back to the original homogeneous peak (i.e., the level of surface antibody was relatively stable over several generations). Serial sorting may provide a more stable population of high secretors. Initial antibody secretion rates followed fluorescence levels (see Table 3); however, the cells that were tested did not possess a range of secretion rates as broad as would be expected for cultures displaying a bimodal intracellular distribution. It is expected that sorting cells that display a wider distribution of intracellular concentrations would result in larger differences in secretion rates of isolated populations. In such a case, low secretors could be more efficiently eliminated from the population, or the highest secretors could be selected. The final IgG secretion rates were low for all

TABLE 3 Antibody Production in Batch Culture by HB124 Hybridoma Cells Sorted Based on Fluorescent Surface Antibody Content[a]

	Surface fluorescence					
	FITC-labeled sort			Bead-labeled sort		
	High	Unsorted	Low	High	Unsorted	Low
% Positive fluorescence on day 4 postseed[b]	34	35	17	23	18	9
Initial specific IgG secretion rate (μg/day) per 10^6 cells[c]	97	95	80	42	31	22
Final specific IgG secretion rate (μg/day) per 10^6 cells[d]	16	18	17	7	4	5
IgG concentration on day 6 postseed (μg/ml)[e]	25.4	25.2	24.6	24.6	25.7	24.9

[a]Removed 4% of cells with highest and lowest surface antibody as determined by surface fluorescence measurements. Surfaces were stained with FITC GAM IgG1 (control) and IgG2a or with anti-IgG antibodies adsorbed onto 0.79-μm–diameter fluorescent latex beads.
[b]Approximately 10,000 cells were analyzed for each sample.
[c]Secretion rates for days 0–1 for FITC-labeled sort and days 0–2 for bead-labeled sort.
[d]Secretion rates for days 1–3 for FITC-labeled sort and days 2–3 for bead-labeled sort.
[e]Error associated with ELISA assay may be as high as 10%.

populations (see Table 3); to benefit from the higher initial secretion rates, cells must be cultured under appropriate conditions.

A similar procedure was followed for a fed-batch suspension culture of sorted cells (data not shown). Four percent of the cells with the highest and lowest surface antibody content were collected, as well as an unsorted population. After 2 days, the sorted cells were seeded in T-flasks. Fresh medium was added periodically, and samples for cell count, viability, and medium antibody concentration were obtained over a 10-day period. Cells that originally had high surface antibody content had higher secretion rates on average than those with low surface antibody; however, the growth rate for the high secretors was low, resulting in lower medium antibody concentrations. Therefore, a trade-off exists between high secretion and high growth rates. Interestingly, high surface antibody fluorescence and corresponding low growth rates were also observed in other experiments. It is possible that a growth phase followed by a production phase could provide more appropriate conditions for increased production by high secretors. Furthermore, cells exhibit higher specific production rates at slowed growth rates, because cells are able to employ more of their metabolic energy for antibody production than for cell growth. Suzuki and Ollis [38] found that monoclonal antibody production could be increased by 50–120% in cultures

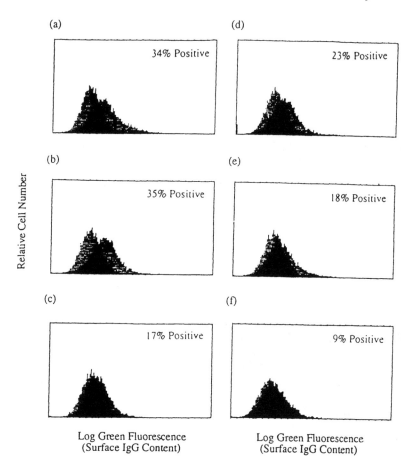

Relative Cell Number

(a) 34% Positive

(b) 35% Positive

(c) 17% Positive

(d) 23% Positive

(e) 18% Positive

(f) 9% Positive

Log Green Fluorescence
(Surface IgG Content)

Log Green Fluorescence
(Surface IgG Content)

FIGURE 8 Surface antibody histograms of sorted populations of HB124 hybridoma cells grown in batch culture that were analyzed on day 4 postseed after staining with FITC-labeled anti-immunoglobulin antibodies. (a, b, and c) the high, unsorted, and low surface antibody populations, respectively, derived from cells stained with FITC-labeled anti-immunoglobulin antibodies; (d, e, and f) the high, unsorted, and low surface antibody populations, respectively, derived from cells stained with antibody-labeled fluorescent beads. Surfaces were stained with FITC GAM IgG1 (negative distribution) and IgG2 (positive distribution). The difference is reported as percentage positive.

treated with growth inhibitors that had no direct effect on antibody synthesis. In a long-term study performed by Ozturk and Palsson [24], losses in antibody productivity were observed as cells were adapted to low serum levels, with improvement in growth rate. Heath et al. [22] showed that IgG secretion rates increased as the percentage of serum in the medium was decreased for cells grown in batch culture. This was thought to be the result of a decreased rate of exponential growth. This study suggested that overgrowth of lower-producing cells may cause loss in culture productivity over time, stressing the importance of reseeding cultures with cells selected to be high producers.

V. CONCLUSIONS

As shown from these experiments, flow cytometry can be a powerful tool to help explain behavior of cells in culture. Surface antibody fluorescence was often correlated with antibody production and was subsequently used as a parameter by which to sort higher- from lower-secreting cells. Such results are important for optimization and maintenance of high-producing hybridoma cultures.

REFERENCES

1. P. J. Leibson, M. R. Loken, S. Panem, and H. Schreiber, Clonal evolution of myeloma cells leads to quantitative changes in immunoglobulin secretion and surface antigen expression, *Proc. Natl. Acad. Sci. USA 76*:2937 (1979).
2. E. Meilhoc, K. D. Wittrup, and J. E. Bailey, Application of flow cytometric measurement of surface IgG in kinetic analysis of monoclonal antibody synthesis and secretion by murine hybridoma cells, *J. Immunol. Methods 121*:167 (1989).
3. M. Dalili and D. F. Ollis, A flow cytometric analysis of hybridoma growth and monoclonal antibody production, *Biotechnol. Bioeng. 36*:64 (1991).
4. P. Marder, R. S. Maciak, R. L. Fouts, R. S. Baker, and J. J. Starling, Selective cloning of hybridoma cells for enhanced immunoglobulin production using flow cytometric cell sorting and automated laser nephelometry, *Cytometry 11*:498 (1990).
5. S. Sen, W.-S. Hu, and F. Srienc. Flow cytometric study of hybridoma cell culture: Correlation between cell surface fluorescence and IgG production rate, *Enzyme Microb. Technol 12*:571 (1990).
6. M. Al-Rubeai, A. N. Emery, and S. Chalder,. Flow cytometric study of cultured mammalian cells, *J. Biotechnol. 19*:67 (1991).
7. M. Leno, O. W. Merten, F. Vuillier, and J. Hache, IgG production in hybridoma batch culture: Kinetics of IgG mRNA, cytoplasmic, secreted- and membrane-bound antibody levels, *J. Biotechnol. 20*:301 (1991).
8. K. L. McKinney, R. Dilwith, and G. Belfort, Manipulation of heterogeneous hybridoma cultures for overproduction of monoclonal antibodies, *Biotechnol. Prog. 7*:445 (1991).
9. M. Al-Rubeai, A. N. Emery, S. Chalder, and D. C. Jan, Specific monoclonal antibody productivity and the cell cycle—comparisons of batch, continuous and perfusion cultures, *Cytotechnology 9*:85 (1992).
10. J. M. Coco-Martin, J. W. Oberink, T.A.M. van der Velden-de Groot, and E. C. Beuvery, The potential of flow cytometric analysis for the characterization of hybridoma cells in suspension cultures, *Cytotechnology 8*:65 (1992).

11. K. L. McKinney, R. Dilwith, and G. Belfort, Optimizing antibody production in batch hybridoma cell culture, *J. Biotechnol.* *40*:31(1995).

12. G. L. Altshuler, R. Dilwith, J. Sowek, and G. Belfort, Hybridoma analysis at cellular level, *Biotechnol. Bioeng. Symp.* *17:*725 (1986).

13. M. Al-Rubeai and A. N. Emery, Mechanisms and kinetics of monoclonal antibody synthesis and secretion in synchronous and asynchronous hybridoma cell cultures, *J. Biotechnol.* *16*:67 (1990).

14. O. T. Ramirez and R. Mutharasan, Cell cycle- and growth phase-dependent variations in size distribution, antibody productivity, and oxygen demand in hybridoma cultures, *Biotechnol. Bioeng.* *36*:839 (1990).

15. S. J. Kromenaker and F. Srienc, Cell-cycle-dependent protein accumulation by producer and nonproducer murine hybridoma cell lines: A population analysis, *Biotechnol. Bioeng.* *38*:665 (1991).

16. P. M. Hayter, N. F. Kirkby, and R. E. Spier, Relationship between hybridoma growth and monoclonal antibody production, *Enzyme Microb. Technol.* *14*:454 (1992).

17. T. I. Linardos, N. Kalogerakis, and L. A. Behie, Cell cycle model for growth rate and death rate in continuous suspension hybridoma cultures, *Biotechnol. Bioeng.* *40*:359 (1992).

18. B. L. Maiorella, Medium design: A new look at "feeding the baby," presented at Cell Culture Engineering II, Engineering Foundation Conferences, Santa Barbara, CA, 1989.

19. S. Reddy and W. M. Miller, Effects of environmental stress on hybridoma antibody production and metabolism, presented at AIChE Annual Meeting, San Francisco, CA, Nov. 1989.

20. S. S. Ozturk and B. O. Palsson, Effect of medium osmolarity on hybridoma growth, metabolism, and antibody production, *Biotechnol. Bioeng.* *37*:989 (1991).

21. K. K. Frame and W.-S. Hu, The loss of antibody productivity in continuous culture of hybridoma cells, *Biotechnol. Bioeng.* *35*:469 (1990).

22. C. Heath, R. Dilwith, and G. Belfort, Methods for increasing monoclonal antibody production in suspension and entrapped cell cultures: Biochemical and flow cytometric analysis as a function of medium serum content, *J. Biotechnol.* *15*:71 (1990).

23. G. M. Lee and B. O. Palsson, Immobilization can improve the stability of hybridoma antibody productivity in serum-free media, *Biotechnol. Bioeng.* *36*:1049 (1990).

24. S. S. Ozturk and B. O. Palsson, Loss of antibody productivity during long-term cultivation of a hybridoma cell line in low serum and serum-free media, *Hybridoma* *9*:167 (1990).

25. G. M. Lee, A. Varma, and B. O. Palsson, Application of population balance model to the loss of hybridoma antibody productivity, *Biotechnol. Prog.* *7*:72 (1991).

26. D. E. Martens, C. D. Gooijer, C.A.M. van der Velden-de Groot, E. C. Beuvery, and J. Tramper, Effect of dilution rate on growth, productivity, cell cycle and size, and shear sensitivity of a hybridoma cell in a continuous culture, *Biotechnol. Bioeng.* *41*:429 (1993).

27. S. E. Merritt and B. O. Palsson, Loss of antibody productivity is highly reproducible in multiple hybridoma subclones, *Biotechnol. Bioeng.* *42*:247 (1993).

28. J. M. Salazar-Kish and C. A. Heath, Comparison of a quadroma and its parent hybridomas in fed batch culture, *J. Biotechnol. 30*:351 (1993).
29. K. L. McKinney, Methods to increase antibody production and improve bioreactor performance for hybridoma cell cultures, Ph.D. thesis, Rensselaer Polytechnic Institute, Troy, NY, 1993.
30. C. Heath, Engineering aspects of improved antibody production by hybridomas, Ph.D. thesis, Rensselaer Polytechnic Institute, Troy, NY, 1988.
31. S. S. Ozturk and B. O. Palsson, Effect of initial cell density on hybridoma growth, metabolism, and monoclonal antibody production, *J. Biotechnol. 16*:259 (1990).
32. T. A. Bibila and M. C. Flickinger, A model of interorganelle monoclonal antibody transport and secretion in mouse hybridoma cells, *Biotechnol. Bioeng. 38*:67 (1991).
33. T. M. Berg, K. Oyaas, and D. W. Levine, Hyperosmolarity and osmoprotection. effects on murine hybridoma cell lines, poster presentation at 5th European Congress on Biotechnology, 1990.
34. S. Reddy, K. D. Bauer, and W. M. Miller, Determination of antibody content in live versus dead hybridoma cells: Analysis of antibody production in osmotically stressed cultures, *Biotechnol. Bioeng. 40*:947 (1992).
35. S. K. W. Oh, P. Vig, F. Chua, W. K. Teo, and M. G. S. Yap, Substantial overproduction of antibodies by applying osmotic pressure and sodium butyrate, *Biotechnol. Bioeng. 42*:601 (1993).
36. K. T. Powell and J. C. Weaver, Gel microdroplets and flow cytometry: Rapid determination of antibody secretion by individual cells within a cell population, *Biotechnology 8*:333 (1990).
37. D. R. Parks, V. M. Bryan, V. T. Oi, and L. A. Herzenberg, Antigen-specific identification and cloning of hybridomas with a fluorescence-activated cell sorter, *Proc. Natl. Acad. Sci. USA 76*:1962 (1979).
38. E. Suzuki and D. F. Ollis, Enhanced antibody production at slowed growth rates: Experimental demonstration and a simple structured model, *Biotechnol. Prog. 6*:231 (1990).

3

Antibody Secretion Assays Using Gel Microdrops and Flow Cytometry

JAMES C. WEAVER
Massachusetts Institute of Technology and One Cell Systems, Inc., Cambridge, Massachusetts

JOHN F. DUNNE AND FOREST GRAY
Syntex Research, Palo Alto, California

KRISTINA G. LAZZARI AND JAMES B. LIN
One Cell Systems, Inc., Cambridge, Massachusetts

I. BACKGROUND

A. Essential Features of Conventional Flow Cytometry Using Suspended Cells

Conventional flow cytometry has become an essential technique for quantitatively assessing cellular subpopulations, is increasingly employed in applications ranging from fundamental biological research to clinical assays [1–3], and now is also of interest in cell culture [4]. Ordinarily suspended cells are analyzed, but occasionally larger objects, such as small multicellular organisms, beads, or gel microdrops (GMDs), are used. The latter provide the basis of the antibody (Ab) secretion assay.

39

The power of flow cytometry lies in the ability to base rapid analysis at the single cell level on multiple parameter measurements. Specifically, flow cytometric measurements use one or two light scatter parameters (typically "side scatter," at 90° from the axis of the illuminating laser, and "forward scatter," at a small angle from an illuminating laser) and several fluorescence emission bands [typically green fluorescence (GF) and red fluorescence (RF) for blue light excitation at 488 nm], but up to five "fluorescence colors" can be used in some flow cytometers. The light scatter signals provide triggering information and semiquantitative cell morphology information, whereas the fluorescence signals usually measure the amount of fluorescent dyes associated with each measured cell. In this sense, conventional flow cytometry usually provides cellular analysis based on cell composition. Relatively nonspecific fluorescent dyes indicate the amount of certain classes of biological materials [e.g., fluorescein isothiocyanate (FITC) for total protein, propidium iodide (PI) for total double-stranded nucleic acids], or highly specific fluorescence-labeling using fluorescence-labeled antibodies or hybridization assays to determine the amount of specific cellular components.

B. Limitations of Conventional Flow Cytometry and Advantages of Gel Microdrops

However, with flow cytometry, only limited functional assay capability ordinarily exists. For example, the use of vital stains can partially assess cell viability, in that the combined functionality of intracellular enzyme activity and membrane integrity are determined. Other examples are transmembrane voltage-sensitive dyes and intracellular pH indicators, both of which respond to general metabolic activity. By combining flow cytometric analysis of such activities with prior exposure to potential inhibitors, such as antibiotics, some assay capability for drug susceptibility is achieved. The general limitation is, however, that only the cell itself is measured by flow cytometry. In contrast, many traditional nonflow cytometric cell function assays fundamentally involve measurement of changes that take place in the extracellular region. For example, determination of the concentration of secreted Ab in the supernatant in microtiter wells, or the measurement of extracellular pH shifts in microbial activity assays, both are based on measurements of the environment external to the cells. Thus, conventional flow cytometry cannot measure Ab secretion. As reviewed briefly in the following, the use of gel microdrops (GMDs) with flow cytometry allows many traditional cell function assays to be microminiaturized, and to be simultaneously combined with cell composition assays. Although many different types of function and composition assays appear possible [5], only a few GMD-based assays have as yet been explored. [5–17]

II. CELL MICROENCAPSULATION TO FORM GEL MICRODROPLETS

A conventional cell suspension is readily converted into a GMD preparation by employing a specialized microdrop maker (CellSys, One Cell Systems, Cambridge, MA; Fig. 1). This involves several emulsification steps (E-1 through E-4), followed by three centrifugation steps (C-1 through C-3). In a typical protocol a 400-µl aliquot of CelGel (One Cell Systems 3% w/v biotin-modified agarose) is liquefied at 90°C for 10 min and cooled to 37°C. The molten CelGel is mixed with 50 µl of fetal bovine serum (FBS) and 50 µl of cell suspension also at 37°C (2×10^7 cells), and then briefly mixed again by vortexing. Immediately thereafter 300–400 µl of the resulting agarose–cell mixture is added to 15 ml of 200 cs dimethylpolysiloxane (37°C), and the preparation is emulsified in the CellSys for 1 min (step E-1) at a blade rotation setting of 2100–2500 rmp at 37°C (step E-2), 1 min at the same setting at 4°C (step E-3), and 10 min at 1100 rmp at 4°C (step E-4). The resulting cooled GMDs are separated from the dimethylpolysiloxane fluid by the following centrifugation protocol: (step C-1) transfer of the GMD–dimethysiloxane suspension into two 15-ml Falcon polyethylene, conical bottom screw-cap tubes; (step C-2) addition of a 7-ml phosphate buffered saline (PBS) under layer to each tube; and (step C-3) centrifugation at 900 g for 5 min. The resulting GMD pellets are transferred to a single clean Falcon 15-ml test tube and resuspended in 13 ml PBS and centrifuged at 500 g for 5 min. This final GMD pellet is then resuspended in either PBS or culture medium in preparation for assay. The overall microencapsulation protocol requires about 60–90 min, and produces approximately 10^7 GMDs, with a mean diameter of 25 µm (about 10^6 contain one or more cells).

An alternative method for forming GMDs is based on a forced disruption of a fluid stream, but this has several important disadvantages. This "vibrating orifice" method uses essentially the same type of fluidics that is the basis for conventional electrostatic sorting. In this method, a liquid jet is forced from a

TABLE 1 Terminology Relating to the Statistical Inoculation of GMDs

Sketch[a]	Symbol	Terminology
◯	$n = 0$	Unoccupied GMD
⊙	$n = 1$	Individually occupied GMD
⊙	$n \geq 1$	Multiply occupied GMD

[a]Dot, individual cell; circle, GMD.
Source: Ref. 5.

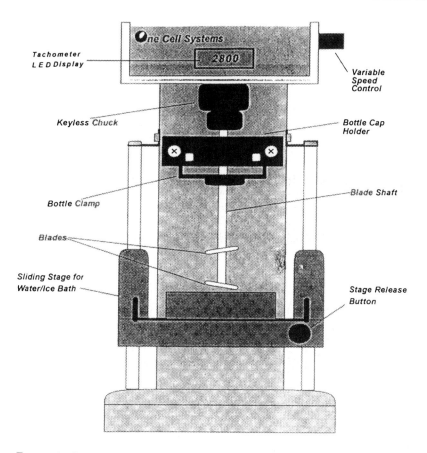

FIGURE 1 Specialized device [CellSys 100 microdrop maker (One Cell Systems, Cambridge, MA)] for forming gel microdrops (GMDs) by microencapsulating suspended cells. A suspension with about 10^6 cells per milliliter of mammalian cells is prepared as the first step in the microencapsulation process. Convenient and reliable microencapsulation is routinely accomplished by using this device which provides biocompatible dispersion of a conventional cell suspension into liquid microdrops, and subsequent temporary cooling, while stirring, to cause gelation of these microdrops to form GMDs. By chosing the proper cell suspension density and average GMD size, approximately 90% of the GMDs are unoccupied, 9% are individually occupied, and fewer than 1% are multiply occupied (see Table 1 for terminology explanation).

small orifice, with the vibration forcing the jet to break up into a parade of liquid microdrops [6]. An advantage is nearly uniform microdrop size, which is often helpful even if strictly unnecessary. The disadvantages include the strict need to filter the solutions and media to avoid clogging the orifice, and also the existence of relatively harsh conditions (high temperature and pressure), which can lead to poor cell viability, even for some relatively robust microorganisms [13]. It is not clear whether this method can be made generally compatible with mammalian cells. For example, the use of air cooling of liquid microdrops involves evaporative loss, a large reduction in microdrop volume and, therefore, significant gel compression. Finally, this type of microdrop formation process is serial. This means that proportionally longer run times (or additional orifices) are required to make larger numbers of microdrops. In contrast, even though filtration before encapsulation is generally useful because it removes debris and particulates, the dispersion method does not absolutely require filtration. It is also carried out at a controlled temperature, with insignificant evaporative loss or gel compression. As a result, cell viability is high, even with mammalian cells [5,15]. Although not yet as frequently used, other gel matrices, such as calcium alginate, can be used, in which gelation occurs by ion exchange, rather than temperature change [6].

A. Size, Permeability, and Strength of Gel Microdroplets

The GMD size distribution has a complicated dependence on the properties of the cell suspension, nonaqueous fluid used for emulsification, and the hydrodynamic dispersion forces caused by the speed and shape of the rotating blade assembly of the CellSys. Typical sizes (GMD diameters) range from about 10 to more than 100 μm, depending on the application. Here, the mean size is about 25 μm, which is appropriate for encapsulating a single mammalian cell, such that the cell is surrounded by a large number of antibody capture sites. For assays involving biochemical activity of single microorganisms, 10- to 20-μm GMDs are appropriate if a rapid result is important [9]. For growth and drug susceptibility assays based on microcolony formation, larger GMDs are appropriate (e.g., larger than 100 μm if mammalian cells are involved). Multiple encapsulation can also be used, forming larger GMDs. For example, double GMDs have been used to locate the inoculum away from the outer regions of the GMDs. This allows more growth to take place before outgrowth occurs, and provides a basis for isolating slow-growing cells from a population of rapidly growing cells [18]. The strength of GMDs is governed mainly by the composition of the gel matrix, but also depends on GMD size, because hydrodynamic shear is often involved. The GMDs are highly permeable and have rapid diffusional response times for exchanging molecules between encapsulated cells and the medium bathing the

TABLE 2 Large GMD Permeabilities Result in Small Diffusion Times

Cell type	Generation time (h)	GMD size(μm)	Response time, $5\tau_D$[a]
Mammalian	12–24	40–100	0.8–5 min
Yeast	1 to several	20–50	0.2–1.3 min
Bacterial	0.3–1	10–25	3–20

[a]Assumes no hindrance effects and $D \approx 10^{-7}$ cm^2 s^{-1} (e.g., myosin, $Mr \approx 500$ kD).
Source: Ref. 5.

GMDs (Table 2). The gel matrix can be considered approximately as a region of unstirred aqueous medium. This is true even for macromolecules, such as fluorescence-labeled antibodies [12] and phycoerythrin-labeled concanavalin A [19]. The effective size cutoff is typically about 500 kDa for the types of agarose usually employed. Nutrients and metabolic wastes are readily exchanged, and immunoassays can be carried out, while retaining cells. Agarose GMDs can be handled much like cells, but are somewhat more fragile, particularly for larger sizes that experience larger shear forces. Thus, with a few modifications the usual manipulations such as pipetting, filtering, centrifuging, and so on, can be used with GMDs.

B. Properties of Gel Microdrops

Here, we emphasize *open GMD*; that is GMDs surrounded by an aqueous medium that allows rapid access to the GMD interior by diffusion. *Closed GMDs* are GMDs surrounded by a nonaqueous fluid; these have been used as the basis of rapid microbial assays [6–9]. The small size and open gel matrix of GMDs make them highly permeable. Although the permeability varies with gel type, for low-melting–temperature agarose, this results in relatively unrestricted and rapid molecular transport for species up to about 500+kDa. This means that cells within GMDs are readily and rapidly exposed to different chemical envi-

TABLE 3 Examples of Physical Manipulation of GMDs

Pipetting	Filtering
Centrifuging	Plating onto petri dishes
Electrical pulsing	Washing
Flow cytometry analysis	FACS (fluorescent-activated cell sorting)

ronments. All that is needed is a change in the chemical composition of the medium in which the GMDs are suspended or bathed. Diffusion rapidly reaches a steady state (see Table 2). Because of this high permeability, cells and micro-colonies within GMDs can be rapidly exposed to new chemical conditions. Likewise, assay protocols involving application of reagents followed by "washing" can be carried out, provided that somewhat longer times are used than for conventional cell suspensions.

C. Number of Cells per Gel Microdrop

A conventional cell suspension contains cells located randomly relative to position. For this reason, if the GMD volumes are significantly larger than the cells, the statistical distribution of cells within GMDs is expected to be random. In such cases, several studies have shown that Poisson statistics adequately describes the distribution of the initial cells for GMDs of different volumes [6,11,13]. Thus, the inoculation of GMDs, microtiter wells, and petri dishes with cells, all are described by Poisson statistics. Recently, however, the use of filter-sterilized agarose with hybridoma cells in 25-μm GMDs resulted in GMDs with initial cells near the center, consistent with the cell nucleating GMD formation (Dunne et al., unpublished results). The limiting case would be a cell with a thin coating of gel. This is nonrandom occupation and also results in fewer multiply occupied GMDs than predicted by Poisson statistics. It is not surprising that basic Poisson statistics should become inadequate as GMD volumes approach cell volumes, because the assumptions underlying Poisson statistics include that of cells having negligible sizes compared with the GMDs.

For initial estimates, however, Poisson statistics is still useful. Mathematically, if the GMD volume, V_{GMD} is sufficiently small, then many GMDs have a high probability of initially containing zero or one cell. This means that the fraction $P(n,\bar{n})$ of GMDs containing n cells depends on both the cell concentration, ρ_{cell}, in the agarose-containing cells suspension from which GMDs are formed, and on the volume of the GMDs. As a consequence, the statistical distribution of initial cells (strictly colony-forming units; CFUs) is quantitatively described by the Poisson formula

$$P(n,\bar{n}) = \frac{\bar{n}^n \, e^{-\bar{n}}}{n!} \tag{1}$$

The quantity $\bar{n}$ is the average number of cells initially contained in a GMD of volume V_{GMD}, and ρ_{cell} is the cell concentration in the agarose-containing suspension from which the GMDs are formed. The average number of cells in a particular-sized GMD is $\bar{n} = \rho_{cell} V_{GMD}$. This means that the probability of unoccupied GMDs (zero initial cells) is $P(0,\bar{n}) = e^{-\bar{n}}$, and the probability of individually occupied GMDs (one initial cell) is $P(1,\bar{n}) = \bar{n}e^{-\bar{n}}$. The occurrence of more than one initial cell in a GMD is governed by chance, so that the probability is

$$P(n > 1, \bar{n}) = 1 - [P(0, \bar{n})] - P(1, \bar{n}) \approx \frac{\bar{n}^2}{2} \tag{2}$$

An acceptable level of "multiple occupation" for many assays is less than 1%, which is achieved if $\bar{n} \approx 0.15$ or lower. Equivalently, V_{GMD} should be less than about $0.15/\rho_{cell}$. The terminology relating to "occupation" of GMDs refers to the number of initial cells present when GMDs are formed, not to the number resulting from subsequent growth (see Table 1). It should also be noted that the formation of GMDs achieves a statistical isolation of cells that is entirely analogous to plating cells. In both cases, the initial cells are separated from the other cells, which is fundamental to providing an ability to use mixed cell populations, or to measure and isolate valuable subpopulations (Fig. 2).

III. GENERAL FEATURES OF GEL MICRODROP BASED ASSAYS

Clonal growth and antibody secretion studies are general types of GMD-based cell function assays that are suited to flow cytometry. The basic idea underlying cellular function assays at the single-cell level is association of a microscopic extracellular region with the cell, such that this extracellular region is physically coupled to the cell while still providing chemical and physical manipulability. In some studies, such as the antibody secretion assay, the extracellular region can be small (i.e., pericellular). This allows optical measurements of both the microscopic extracellular region and the cell by flow cytometry, and sorting interventions using a fluorescence-activated cell sorter (FACS). Briefly, three types of cellular function assays have been emphasized: (a) microbial activity in closed GMDs, (b) clonal growth assays in open GMDs, and (c) antibody secretion assays in open GMDs. Here, we consider the present status of the latter. A significant improvement is based on biotinylated agarose, which can be used to provide an easily generated specific affinity matrix [20]. This class of assays is the main topic of this chapter, but should be considered within the context of the general combined cellular function or composition assays that can be approached using GMDs and flow cytometry.

A. Essential Features of Secretion Assay Protocol

Encapsulated cells are assayed for secretion by generating a target-specific capture matrix in the GMDs after encapsulation, and then labeling the captured molecule in a "sandwich" format. The biotin sites in the agarose of the formed

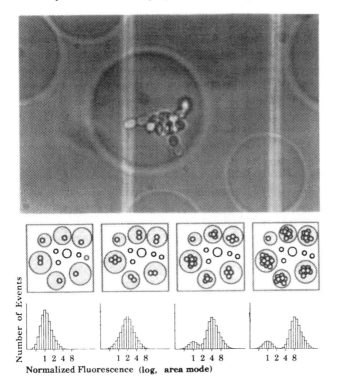

FIGURE 2 Schematic illustration of culture of cells within GMDs such that initial single cells lead to microcolony formation. Subsequent measurement of cell–microcolony biomass allows the number of cells in each GMD to be determined. For cells that grow within a gel matrix, this allows cytoxicity and other cell growth-based assays to be carried out at the individual cell level, with analysis by flow cytometry. For antibody secretion assays the same measurement capability allows determination of the number of secreting cells in each GMD, so that cell secretion normalized to cell number can be measured. This provides a basis for using very long incubations with the secretion assay, as growth and Ab secretion can be determined simultaneously by using two different fluorescent stains (one for cell, e.g., a membrane stain, and the other as the reporter Ab in the secretion assay). (From Ref. 15.)

GMDs are linked to capture molecules (usually Abs) by sequential incubation of the GMDs with a streptavidin "bridge," and then with a biotinylated capture Ab (see Fig. 4). All of these incubations are carried out at 4–37°C as needed, and the GMDs are washed in PBS between incubations to remove excess reagents. The encapsulated cells are then allowed to secrete into the gel matrix, which selectively retains the target molecules while allowing other molecules to diffuse out into the surrounding aqueous medium. Next, GMDs are incubated with a fluorescence-labeled reporter Ab that binds to the captured molecule, resulting in the GMD having reporter fluorescence in proportion to the number of captured target molecules. As in all assays and optical measurements, the issues of nonspecific staining and fluorescence background must be considered, but in GMDs they can be reduced to very low levels (see Fig. 5).

B. Flow Cytometry and Sorting of Gel Microdrops

The analysis and sorting of entities larger than cells is a straightforward extension of ordinary flow cytometry, provided that basic geometric and hydrodynamic issues are considered [1–3]. Clearly the fluidics must accommodate the GMDs, and in practice, the most significant limitation is the need to use a larger sorting orifice, because the GMD diameter should not exceed about 0.3 of the orifice diameter [21, 22]. This means that lower orifice frequencies are needed, and therefore, the throughput rate is smaller than for conventional mammalian cell suspensions. For 25-μm GMDs, this means that a few hundred occupied GMDs s^{-1} (of order 10^3 s^{-1} total) are processed. Gating on occupied GMDs allows analytical attention to be devoted to the GMDs most immediately relevant to the secretion assay. As discussed later, however, measurement of unoccupied GMDs can be included to detect the onset of cross-talk.

There are other minor limitations or adjustments that should be considered in analyzing GMDs, particularly the larger sizes (e.g., > 50 μm). Most flow cytometers are designed and set up for conventional mammalian cell analysis. This means that the focal volume of the fluorescence excitation light is smaller than the larger GMDs, with the result that the GMDs are nonuniformly illuminated. This means less fluorescence excitation light reaching the regions of the GMD away from the flow axis. If all GMDs had a single cell at their centers, this would not be significant. However, inoculation of GMDs appears to involve random locations of cells within them. Consequently, the same cell, with its surrounding cloud of captured Ab, could generate different-sized fluorescence secretion signals, which places a limitation on the accuracy of the secretion assay. This limitation is not intrinsic; it is solely a consequence of the design characteristics of commercially available flow cytometers. Another minor limitation relates to the electronics in most flow cytometers. If a trigger signal (e.g., light scatter) results in a gated measurement time of a few microseconds, this is long enough to acquire the complete signal from a 10- to 15-μm–diameter mammalian cell, but may stop acquiring optical signal before a much larger GMD has passed through the optical interrogation region. This means that the "captured Ab" can be incompletely measured. This too is not intrinsic, but may be a limitation in some flow cytometers. Finally, in those flow cytometers that provide a choice, comparison of "peak" and "area" measurement modes can be used, with area less susceptible to error caused by a first-encountered small peak being measured instead of a subsequent large peak.

IV. GENERAL ASPECTS OF MOLECULAR SECRETION ASSAYS USING GEL MICRODROPS AND FLOW CYTOMETRY

The basic idea underlying the GMD-based assay for antibody secretion by individual cells is illustrated in Figure 3 [12, 23]. As indicated there, this idea is to capture secreted molecules within the gel matrix of the GMD that contains the secreting cell. Conceptually, the following steps can be distinguished: (a) A molecule is secreted by a cell that is entrapped within the highly permeable gel matrix; (b) the molecule diffuses within the gel matrix; (c) the molecule encounters a capture site; (d) with some probability p, the molecule is captured, or with probability $1 - p$ is not captured, and diffuses further; (e) the molecule is finally captured within the originating GMD or escapes into the surrounding medium; (f) escaped molecules experience several general fates: they remain in solution, or are captured at binding sites in other, nonoriginating GMDs, or in scavenging particles that have been provided in the medium external to the GMDs. The probability of returning to the originating GMD is small, and can be neglected. After an incubation during which secretion takes place, a fraction of the secreted molecules have been captured within their originating GMDs, and the remainder have reached other, nonoriginating GMDs (this constitutes "chemical cross-talk"), or has been scavenged.

A. Cross-Talk Is Not a Real Problem

Early experiments revealed both the existence of cross-talk and the basis for rendering it insignificant [12]. For practical applications, an assay system does not have to work perfectly, and that is true here. In principle, at least a few Ab molecules will always escape their originating GMD. The real question is: What is the fate of escaped Ab molecules, and how do they actually affect the assay result? Most applications of the Ab secretion assay seek to find hybridoma cells that are high producers of an Ab; low producers are generally not wanted.

First consider short secretion incubation times (a few minutes to a few hours), as these have been found adequate [12, 20, 24, 25]. Over the time of a short incubation most of a cell's Ab production is captured in the pericellular region of the gel matrix. This has been confirmed by fluorescence microscopy after reporter Ab has been provided and the GMDs then washed: a bright cloud of reporter fluorescence, localized around the cell, is seen. This means that only

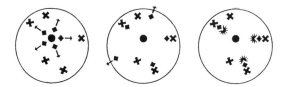

FIGURE 3 Schematic illustration of capture and assay of secreted molecules within GMDs. The basic idea is that secreted molecules are captured at sites provided within the GMD that contains the secreting cell. This is accomplished by providing a large number of capture sites within the gel matrix of GMDs. In addition, the gel matrix is highly permeable, such that even fluorescence-labeled reported Ab can be subsequently introduced from the medium in which the GMDs are bathed. This provides the basis for a quantitative immunoassay of the captured molecules. The initial version of the antibody secretion assay utilized coencapsulated microbeads with surface attached capture Ab. (From Ref. 12.)

a small fraction of the secreted Ab escapes to contaminate unoccupied GMDs and GMDs containing low secretors. Here, the worst that occurs is a slight increase in reporter signal from unoccupied GMDs and GMDs occupied by low producers. As shown in the following, the "contamination" of a GMD containing a high secretor ("H cell") by a low secretor ("L cell") is a more significant issue than cross-talk.

B. Probability of Finding One High Secretor and One Low Secretor in a Gel Microdrop

The use of GMDs to isolate H cells is fundamentally an enrichment process, and one should not be expected to find rarely occurring H cells from a population of many L cells in a single step. The main limitation is the frequency of occurrence of L cells in GMDs that contain H cells. Usually light scatter or fluorescence cell staining can be used to gate out multiply occupied GMDs. Furthermore, in some cases, GMD formation favors singly occupied GMDs, so that the $n = 1$ case has a larger frequency of occurrence than predicted by Poisson statistics. The following estimate based on Poisson statistics is used to give an idea of what would occur in a worst case in which Poisson statistics governs GMD occupation and multiply occupied GMDs are not gated out.

Here we are interested in the probability of finding exactly one L cell and one H cell in the same GMD. We can write this probability formally as

$$P(n_L=1, n_H=1) = \frac{1}{2} P(1, \bar{n}_L) P(1, \bar{n}_H) = \bar{n}_L \bar{n}_H \, e^{-(\bar{n}_L + \bar{n}_H)} = \bar{n}_L \bar{n}_H \, e^{-\bar{n}} \quad (3)$$

which we have written as the product of the two independent probabilities of separately finding one L cell in a singly occupied GMD and one H cell in the singly occupied GMD. This can occur in two ways: H + L or L + H, but we do not care about the order, and this means we divide by two, hence, the factor of 1/2. Here $\bar{n} = \bar{n}_L + \bar{n}_H = (\rho_L + \rho_H)\, V_{GMD}$. We can confirm that this is correct by considering the case in which both cells are actually the same type (e.g., let L → H). This should give $P(n_L=1, n_H=1) \to P(2, \bar{n})$. Indeed, substitution gives Eq. (1) with $n = 2$, the case of ordinary double occupancy.

As an illustration of this issues, we consider the realistic case in which the total (overall) occupation of GMDs is 15%. This means $\bar{n} = 0.15$. Suppose further that the H cells are only a fraction $F_H = 0.03$ (3%) of the total population. This means that $\bar{n}_H = F_H\, \bar{n} = 0.03\bar{n} = 4.5 \times 10^{-3}$ and $\bar{n}_L = [1 - F_H]\, \bar{n} = 0.97\bar{n} = 0.146$. The probability of having a doubly occupied GMD with one L and one H, therefore, is

$$P(n_L=1, n_H=1) = \bar{n}_L\, \bar{n}_H\, e^{-\bar{n}} = (0.146)(4.5 \times 10^{-3})\, e^{-0.15} \approx 6 \times 10^{-4} \qquad (4)$$

In this case about 2 GMDs in 1000 will have one L and one H. Therefore, if Poisson statistics provides a reasonable description, it will be unrealistic to decrease this "mixed double occupation" significantly. If Poisson statistics do not adequately describe the inoculation of GMDs, it is usually because cells are "sticky" (not a significant issue with hybridoma cells), or because cells occupy a significant fraction of a GMD volume. If this occurs, there are fewer cells per GMD than predicted by Poisson statistics, so the estimate of Eq. (4) describes a worst case. This example shows that "contamination" by L cells will usually be more of an issue than chemical cross-talk. A productive approach to this issue is to recognize that GMD-based isolation is an enrichment process. This means that a great concentration of H cells can be accomplished, but that GMD-based isolation should not be considered a single-step process that will isolate very rare H cells. Instead, if very rare H cells are valuable, the process should be used two or more times, with significant enrichment obtained with each use. Furthermore, another strategy is the use of cell stains to determine occupancy, so that only GMDs with single cells are used in the analysis leading to cell isolation.

C. Use of An Immunoassay Within Gel Microdrops

The latent signal consisting of the captured molecules can now be made visible (literally), by employing a fluorescence immunoassay that gives each GMD a fluorescence signal that is proportional to the number of captured molecules. This is readily accomplished by supplying a fluorescence-labeled reporter Ab in the aqueous medium containing the GMDs, allowing the reporter Ab to diffuse into the gel, where it binds to captured secreted molecules. The result is a "fluorescent cloud" near the secreting cell, which can be observed using fluorescence microscopy. For large numbers of quantitative measurements, flow cytometry is used to reveal the amount of reporter fluorescence in each analyzed GMD. Variations of this strategy are clearly possible. For example, the use of a non-fluorescence-labeled mouse Ab, followed by introduction of a goat-antimouse Ab with a fluorescence label, provides straightforward flexibility. Once a fluorescence signal has been developed, the GMDs can be analyzed by flow cytometry. To accomplish cell isolation based on secretion, FACS is employed to sort GMDs based on the secretion fluorescence signal (paying attention to the larger GMD size), and viable cells are recovered by outgrowth or enzymatic digestion of the gel.

V. INTERNAL CONTROL PROVIDED BY UNOCCUPIED GEL MICRODROPS

Unoccupied GMDs are actually useful, provided they can be readily identified. For example, the problem of chemical cross-talk can be reduced if there is a large pool of capture sites available, as when uncaptured molecules are distributed among these sites. This means that each GMD is contaminated by only a small fraction of the cross-talk. Also, because the unoccupied GMDs are, by definition, not sources of the secreted product, the fluorescence level of these GMDs provide a baseline for the assay. The GMDs with greater than baseline fluorescence contain the secreting cells.

A. Advantages of Identifying the Unoccupied Gel Microdrops

Successful quantitative determination of secretion by individual cells within a cell population requires that none of the GMDs saturate their capture sites. The problem of saturation is complicated because the GMD size is usually not tightly controlled. This means that saturation depends on two independent variables: (a) individual cell secretion rate and (b) GMD capture capacity. In this context, unoccupied GMDs serve as monitors of supersaturation; as long as they remain nonfluorescent, the overall assay system determines individual cell secretion independently of GMD size. The GMDs with a volumetric fluorescence label of the gel matrix can be measured using the volume fluorescence as a trigger [18]. In this case, the trigger signal slightly precedes the secretion signal, which is a cloud of another fluorescence color.

B. Use of Coentrapped Beads with Surface Capture Sites

The initial approach to providing a large number of capture sites within the gel matrix of a GMD involved coentrapment of "capture beads" that had a surface coating of binding sites to capture secreted molecules [12, 23]. For this, hybridoma cells were used to demonstrate basic feasibility of the secretion assay. The capture sites were provided by coentrapping approximately 4×10^3 0.8-μm polystyrene beads coated with goat–anti-(mouse IgG) Ab within 80-μm agarose GMDs. Following an 11-h secretion incubation, the immunoassay was carried out, followed by flow cytometry. This initial study demonstrated that secreting cells could be readily distinguished, using both microscopy and flow cytometry. With flow cytometry, the several thousand beads per GMD provided a large light scatter signal, which can be used to trigger fluorescence measurements of the captured molecules and, also, if desired, cell membrane fluorescence signals that allow the number of cells per GMD to be determined.

Although the use of beads to provide capture sites worked in some cases, in others, there was difficulty in isolating viable cells. For example, beads were sometimes phagocytosed by the cells, which then often resulted in dead cells containing many beads. In addition, the large, bead-scatter signal precluded using scatter alone as a basis for distinguishing cell-occupied GMDs. This problem emphasized the need to find a nonbead format for providing capture sites.

C. Improved Molecular Capture Using Biotinylated Agarose

The use of biotinylated agarose (One Cell Systems), admixed with an unmodified agarose, now provides a general approach to creating optically clear GMDs with a high level of capture sites. This important advance was first reported by Gray et al. [20], and subsequently, confirmed by others [25]. As shown in Figure 4, avidin is used to attach biotin-agarose to biotin-Ab, which creates specific capture sites. In the presence of a secreting cell, the secreted molecules (here Ab) diffuse within the gel matrix and are efficiently captured by the high-volume density of capture sites. Subsequent exposure of the GMDs to a fluorescence-labeled reporter Ab results in diffusion into the GMDs, and, thereby, a fluorescence marking of the captured secreted molecules. The overall result is a fluorescence measurement that is proportional to the accumulated secreted molecule.

Agarose GMDs without capture beads are generally optically clear. As a result, cell-based light scatter dominates the scatter signal, which allows occupied GMDs to be distinguished from unoccupied GMDs without needing an additional fluorescence measurement (e.g., a nonspecific membrane stain) to designate the presence of cells. Unoccupied GMDs serve as internal negative controls, as they are exposed to all of the reagents and experience all of the incubation conditions as the occupied GMDs (see Fig. 6).

D. Number of Capture Sites in Biotinylated-Agarose Gel Microdrops

A very large number of capture sites can be provided within the gel matrix (CelGel, One Cell Systems) while retaining a high permeability and optical

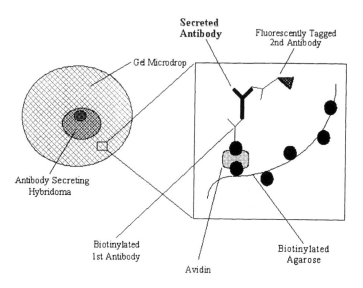

FIGURE 4 Illustration of improved capture of secreted molecules using biotinylated agarose, rather than coencapsulated microbeads. This version greatly increases the number of capture sites, and avoids problems associated with bead endocytosis and with large light scatter from all GMDs, whether or not occupied by cells [20].

clarity. For example, a 30-μm GMD has close to 3×10^{12} capture sites. This and the relatively small degree of nonspecific staining provide the basis for the large dynamic range of fluorescence signals caused by capture of the analyte Ab.

E. Detection of Occupied Gel Microdrops by Flow Cytometry

Formation of an excess of unoccupied GMDs is done to ensure a reasonably high probability of individual cell occupation. As long as occupation of GMDs is random, Poisson statistics governs the distribution of initial cells. Typically, an average occupation of $\bar{n} \approx 0.15$ is used, so that the ratio, R of singly occupied to multiply occupied GMDs is

$$R = \frac{P(1,\bar{n})}{P(n>1,\bar{n})} \approx \frac{2e^{-\bar{n}}}{\bar{n}} \approx 11 \quad \text{if} \quad \bar{n} = 0.15 \tag{5}$$

This means that about 8% (1/[11 + 1]) of the occupied GMDs contain two or more cells. These multiply occupied drops can be easily distinguished by using a vital fluorescent stain, provided that the single cell distribution of fluorescence is

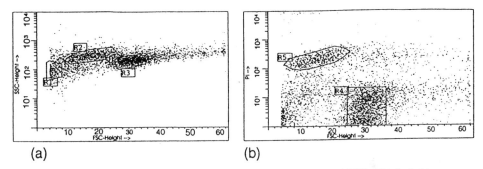

(a)   (b)

FIGURE 5 Light scatter and fluorescence measurements of TIB 114 hybridoma cells microencapsulated in biotinylated 2.5% agarose GMDs. Here, the reporter Ab was PE-labeled anti-IgG Ab (size of approximately 400 kDa), which permeated the gel matrix with insignificant hindrance. (a) SSC (side scatter) versus FSC (forward scatter) for 10^4 GMDs. Three rectangular regions (R1, R2 and R3) and three nonrectangular regions are delineated by solid boundaries. R1 consists of unoccupied (empty) GMDs, R2 consists of occupied GMDs with dead cells, and R3 consists of individually occupied GMDs with viable cells. (b) FSC versus PI-fluorescence of GMDs. R4 contains negative (nonsecreting), viable cells. R5 consists of PI-stained GMDs, which is due to dead cells with compromised membranes allowing entry of PI and intense staining of double-stranded nucleic acids. (From Ref. 25.)

not too large. The issue of single versus multiple occupation is solely dependent on $\bar{n}$ if Poisson statistics governs the distribution. As noted earlier, however, nonrandom, nearly centered cells have been observed in 25-μm GMDs formed from filter-sterilized agarose (Dunne et al., unpublished). Even if the many unoccupied drops are removed by density gradient or other methods, it is still desirable to distinguish single from multiply occupied GMDs, to avoid interpreting the secretion of two cells as that of one cell secreting at a higher rate. However, as long as the GMD-secretion assay is considered as an enrichment process, even some error arising from incorrect isolation of multiply occupied GMDs would not prevent overall success, as only the enrichment factor would be diminished.

With this in mind, the first step is usually the detection of occupied GMDs in the presence of many more unoccupied GMDs. Both light scatter (no staining needed) and fluorescence can be used. One example is shown in Figure 6, which compares light scatter with the fluorescence of a vital stain BCECF (2',7'-bis-(2-carboxyethyl)-5-(and-6)-carboxyfluorescein). Another is shown in Figure 5, which uses light scatter and also propidium iodide (PI) staining to distinguish

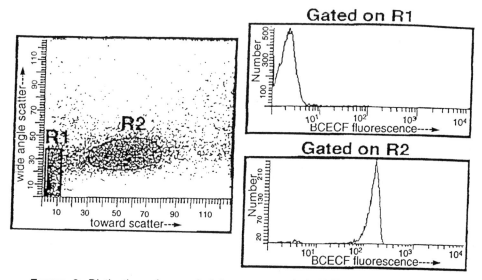

FIGURE 6 Distinction of occupied from unoccupied GMDs. A preparation of GMDs containing hybridoma cells was stained with GF (green fluorescence) using BCECF (Molecular Probes, Eugene, OR) to stain viable cells. The forward and wide angle scatter dot plot revealed two main populations, R_1 and R_2. By also determining that GF region R_1 did not contain viable cells, whereas nearly all the events in R_2 stain brightly with BCECF. This indicates the presence of viable cells within the GMDs. Separate sorting experiments (data not shown) showed that R_2 contained primarily singly occupied GMDs, and events to the right of R_2 were multiply occupied GMDs.

and exclude dead cells in GMDs. If the light scatter approach is employed, a single fluorescence "color" is sufficient for finding secretors.

IV. ANALYSIS AND SORTING OF GEL MICRODROPS CONTAINING HIGH SECRETORS

Biotinylated agarose GMDs have been used to assay and isolate high secretors from populations of several hybridoma cultures. Viable cells were encapsulated in GMDs, sorted (flow-cloned), and then recovered by using agarase. BVD 624.G2.3 is a rat–mouse hybridoma that has been multiply subcloned and productive for several years [26]. The secreted Ab was captured by a biotinylated goat–antimouse immunoglobulin Ab, and used a fluoresceinated goat–antimouse immunoglobulin reporter. This combination of immunoreagents is convenient and appears to have broad utility. However, more specific capture-reporter constructions can also be used. [20, 24] Figure 7 compares the scatter-gated fluorescence distribution of the singly occupied GMDs using cells before GMD-based isolation with the culture that was grown up after sorting the highest 10% of the GMDs based on the secretion signal. The cells from this sorted subpopulation were cloned using an automatic cell deposition device, and four of these clones were characterized by Ab production using conventional microtiter well methods. Specifically, ELISA of the supernatant of these clones revealed that three produced Ab at rates more than eight times that of the parent, but the other had by then ceased to produce detectable Ab.

A. Enumeration of Secreting Cells

If the goal is to determine how many cells within a population are secreting within a particular range, then if the selected range is large, relatively poor secretors need their Ab secretion to be detected in the presence of high secretors. A potential problem is that the good secretors could saturate their GMDs, and by chemical cross-talk contaminate other GMDs. However, a typical preparation has almost tenfold more unoccupied GMDs than cell-containing GMDs. With good mixing this means that this contamination is homogeneous and significantly diluted within the GMD preparation. By identifying the unoccupied GMDs and monitoring their rising contamination level, incubation can be prolonged until the unoccupied GMDs approach the brightness of the dimmest GMDs with secreting cells (Fig. 8).

B. Recovery of Viable Cells from Sorted Gel Microdrops

Two general strategies have proved successful. Cells can be grown out of sorted GMDs, or the agarose matrix can be enzymatically digested, or both. Simple outgrowth is appropriate if the GMDs are relatively small, the gel concentration low, and (which must be determined empirically) the cells are capable of growing in that particular agarose matrix. Currently, it appears best to use agarase digestion. In a typical digestion protocol the enzyme is added in solution to a preparation of GMDs at 37°C, and the agarose matrix degraded without significant loss of cell viability for most cell types yet used.

C. Secretion Incubation Time and Temporal Variation in Secretion Rates

The secretion incubation time is one of the most important variables in the overall analysis-isolation process. For high secretors the incubation times can be short, ranging from about 15 min to a few hours. Essentially all GMDs will contain low levels of secreted Ab at the instant of dispersion in the CellSys100,

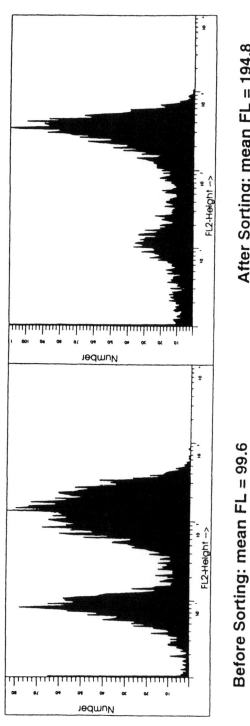

Before Sorting: mean FL = 99.6   **After Sorting: mean FL = 194.8**

FIGURE 7 Scatter-gated reporter fluorescence distribution from singly occupied GMDs made with the unsorted hybridoma cells (left) and GMDs containing cells that were grown up after sorting the top 10% relative to reporter fluorescence (right).

because secretion will have occurred while the cells were in a common suspension. Once separated by GMD formation, however, there is generally rapid localized accumulation of captured Ab and, therefore, reporter fluorescence. If there is significant temporal variation in secretion over longer times, then misleading results might be obtained. However, the real proof of success is isolation of high secretors (H cells) followed by subsequent culture and demonstration that they are stable high secretors.

V. FUTURE PROSPECTS

In addition to directly addressing the problem of measuring Ab secretion at the individual cell level so that "high secretors" can be isolated on a quantitative

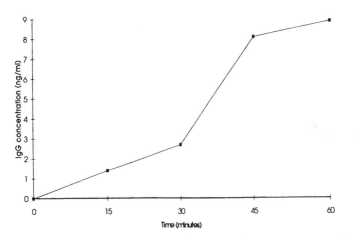

FIGURE 8 Time course of the average PE-fluorescence secretion signal due to captured Ab molecules from TIB 114 myeloma cells. Time zero corresponds to the beginning of a 37°C incubation of GMDs, during which secreted Ab was captured. At each time point, 10^4 GMDs were analyzed by flow cytometry. The rapid accumulation of sufficient captured Ab for fluorescence measurement within minutes is consistent with typical antibody secretion rates and the high sensitivity measurement response demonstrated in Figure 7. Given the speed at which Ab secretion signal accumulated, it was possible to generally use incubation times of 60 min or less in the isolation of the highly secreting subpopulation. This has the additional advantage of minimizing escape of secreted Ab from the originating GMDs, and the resultant contamination of nonoriginating GMDs with secreted Ab (cross-talk between GMDs). Here the Ab secretion signal (PE-fluorescence) was converted into an equivalent IgG concentration (C_{soln}) supplied externally to the particular GMD preparation. $\bar{V}_{GMD}$ is the average GMD volume, V_{soln} the volume of the supplied IgG solution and F_{cal} is a fluorescence calibration factor such that the average GMD fluorescence is $\bar{F} = F_{cal}\bar{N}_{IgG3} = F_{cal}[V_{soln}/\bar{V}_{GMD}]$.

secretion rate per cell basis, rapid determination of the secretion of specific molecules by individual members of a cell population should find many other applications in biotechnology, basic biological research, and medicine. The secretion-capture assay is not limited to natural secretion: stimulated release of intracellular molecules by nonspecific agents, such as electric fields that cause electroporation, should be possible. It has already been demonstrated that cells within agarose GMDs can be electroporated, as indicated by their uptake of a charged fluorescent molecule, and that the viable subpopulation of encapsulated, electrically pulsed cells can be determined. For this, viability is determined by using an incubation corresponding to two or three generation times, and then using a second fluorescent molecule to stain the resulting microcolonies (viable cells) and single cells (nongrowing and presumed dead) [19]. Many other variations on the basic secretion assay are possible. For example, the use of a first fluorescent molecule for the reporter Ab label and a second fluorescent molecule for determining the size of a microcolony provides the basis for long-term growth-secretion assays in which the accumulated secretion is normalized to the number of cells present. This and many other possibilities will undoubtedly be explored by imaginative users of GMD-based assays.

ACKNOWLEDGMENTS

We thank S. J. Sullivan, C. Ryan, E. O'Lear, P. McGrath, B. Goguen, and E. A. Gift for many stimulating discussions. This work is partially supported by NIH SBIR Grant 1 R43GM50042 to One Cell Systems.

REFERENCES

1. E. R. Melamed, T. Lindmo, and M. L. Mendelsohn, in Flow Cytometry and Sorting, 2nd ed., Wiley-Liss, New York, 1990.
2. A. L. Givan, in *Flow Cytometry: First Principles,* Wiley-Liss, New York, 1992.
3. H. M. Shapiro, *Practical Flow Cytometry*, 3rd ed., A. R. Liss, New York, 1994.
4. M. Al-Rubeai, and A. N. Emery, Flow cytometry in animal cell culture, *BioTechnology 11*:572–579 (1993).
5. J. C. Weaver, J. G. Bliss, G. I. Harrison, K. T. Powell, and G. B. Williams, Microdrop technology: A general method for separating cells by function and composition, *Methods 2*:234–247 (1991).
6. J. C. Weaver, P. E. Seissler, S. A. Threefoot, J. W. Lorenz, T. Huie, R. Rodrigues, and A. M. Klibanov, Microbiological measurements by immobilization of cells within small volume elements, *Ann. N.Y. Acad. Sci. 434*:363–372 (1984).
7. J. C. Weaver, Gel microdroplets for microbial measurement and screening: Basic principles, *Biotechnol. Bioeng. Symp. 17*:185–195 (1986).

8. G. B. Williams, S. A. Threefoot, J. W. Lorenz, J. G. Bliss, J. C. Weaver, A. L. Demain, and A. M. Klibanov, Rapid detection of *E. coli* immobilized in gel microdroplets, *Ann. N. Y. Acad. Sci. 501*:350–353 (1987).

9. J. C. Weaver, G. B. Williams, A. M. Klibanov, and A. L. Demain, Gel microdroplets: Rapid detection and enumeration of individual microorganisms by their metabolic activity, *Biotechnology 6*:1084–1089 (1988).

10. A. Rosenbluth, R. Nir, E. Sahar, and E. Rosenberg, Cell density dependent lysis and sporulation of *Myxococcus xanthus* in agarose microbeads, *J. Bacteriol. 171*:4923–4929 (1989).

11. G. B. Williams, J. C. Weaver, and A. L. Demain, Rapid microbial detection and enumeration using gel microdroplets: Colorimetric versus fluorescent indicator systems, *J. Clin. Microbiol. 28*:1002–1008 (1990).

12. K. T. Powell and J. C. Weaver, Gel microdroplets and flow cytometry: Rapid determination of antibody secretion by individual cells within a cell population, *Biotechnology 8*:333–337 (1990).

13. R. Nir, R. Lamed, L. Gueta, and E. Sahar, Single-cell entrapment and microcolony development within uniform microspheres amenable to flow cytometry, *Appl. Environ. Microbiol. 56*:2870–2875 (1990).

14. R. Nir, Y. Yisraeli, R. Lamed, and E. Sahar, Sorting of viable bacteria and yeasts according to ß-galactosidase activity, *Appl. Environ. Microbiol. 56*:3861–3866 (1990).

15. J. C. Weaver, J. G. Bliss, K. T. Powell, G. I. Harrison, and G. B. Williams, Rapid clonal growth measurements at the single-cell level: Gel microdroplets and flow cytometry, *Biotechnology 9*:873–877 (1991).

16. E. A. Gift and J. C. Weaver, Cell survival following electroporation: Quantitative assessment using large numbers of microcolonies, *Electricity and Magnetism in Biology and Medicine* (Blank, ed.), San Francisco Press, 1993, pp. 147–150.

17. E. Sahar, R. Nir, and R. Lamed, Flow cytometric analysis of entire microcolonies, *Cytometry 15*:213–221 (1994).

18. E. A. Gift, H. Park, G. A. Pardis, A. L. Demain, and J. C. Weaver, FACS-based isolation of slow growing microorganisms: A yeast model system involving double encapsulation of yeast in gel microdrops (in preparation).

19. E. A. Gift, A method for measuring electroporative uptake and cell growth using gel microdrops and flow cytometry, M.S. thesis, MIT, 1993.

20. F. Gray, J. S. Kenney, and J. F. Dunne, Secretion capture and report web: Use of affinity derived agarose microdroplets for the selection of hybridoma cells, *J. Immunol. Methods 182*:155–163 (1995).

21. J. H. Jett and R. G. Alexander, Droplet sorting of large particles, *Cytometry 6*:484–486 (1985).

22. R. T. Stoval, The influence of particles on jet breakoff, *J. Histochem. Cytochem. 25*:813–820 (1977).

23. K. T. Powell, Mammalian cell clonal growth and secretion measurements using gel microdroplets and flow cytometry, Ph.D. thesis, MIT, 1989.

24. J. S. Kenney, F. Gray, M.-H. Ancel, and J. F. Dunne, Production of monoclonal antibodies using a secretion capture and report web *Bio/Technology 13*:787–790 (1995).

25. J. B. Lin, K. Lazzari, G. A. Paradis, J. C. Weaver, and S. S. Sullivan, Rapid analysis and isolation of hybridoma cells for antibody secretion using gel microdrops (GMDs) and fluorescent activated cell sorting (FACS), (in preparation).

26. R. L. Coffman, B. W. Seymour, S. Hudak, J. Jackson, and D. Rennick, Antibody to interleukin 5 inhibits helminth-induced eosinophilia in mice, *Science 245*:308–310 (1989).

4

High-Resolution Cell Cycle Analysis of Cell Cycle-Regulated Gene Expression

MANFRED KUBBIES AND BERNHARD GOLLER
Boehringer Mannheim GmbH, Penzberg, Germany

GUENTER GIESE
Max-Planck Institute of Cell Biology, Ladenburg, Germany

I. INTRODUCTION

Cellular integrity, physiological homeostasis, identical replication of the genome, and cell division are regulated in a complex network of interactions of different kinds of molecules. These encompass not only proteins, but also lipids and derivatives thereof, nucleic acids, and low molecular weight components. The function and interaction of the various cellular components are regulated essentially in two different ways:

1. The physiological action might be constitutive, as it is, for example, for the energy-supplying biochemical pathway. Most of these housekeeping genes and proteins are expressed constitutively during the cell cycle. This is also valid for many luxury proteins expressed in differentiated cells.

2. The physiological function of genes or proteins might be necessary only during a distinct period of the cellular life cycle. Most of the molecules necessary for the replication of the genome (i.e., DNA synthesis proteins, or regulatory molecules like cyclins) are expressed, or at least regulated, in a cell cycle-dependent fashion.

The analysis of cell cycle-regulated gene expression has as a prerequisite either single-cell analytical tools or isolation of synchronous, physiologically homogeneous cell populations. Many or most of the molecular analytical techniques using cellular homogenates do not reveal the population heterogeneity that exists even in permanent cell lines and clones derived therefrom:

1. *Cell cycle-specific gene expression:* Asynchronously or partially synchronized cells are in different compartments of the cell cycle, or at least traverse different time or checkpoints in identical compartments (e. g., early and late G_1-phase cells).
2. *Heterogeneity of expression:* Identical gene expression values may be obtained either by all cells exhibiting low, or by a few cells exhibiting high, gene expression rates.
3. *Multiple independent kinetic processes:* (a) Induction of a physiological process with increasing rates of expression and, in parallel, (b) cell cycle progression (delayed analysis after induction).
4. *Cellular amount or concentration of a gene product:* Is it the difference of expression owing to increased synthesis, or is it simply determined by differences in cell size?

There are some indirect approaches to solving these analytical problems, including purification techniques such as elutriation used to obtain homogeneous cell populations as starting populations for molecular analysis [1], stathmokinetic methods of enrichment by inhibition of proliferating cell populations in distinct cell cycle compartments [2], or growth factor addition or removal [3]. These are frequently used approaches. However, apart from the continuing problem of the detection limits of low quantities of gene expression or gene products in single cells, for many approaches flow cytometry offers the optimal tool for solving most of the aforementioned analytical difficulties. Flow cytometry has the unique property of recording quantities of biomolecules in individual cells at high speed in a multiparameter fashion [4–6]. This technique, therefore, offers unprecedented insights into the heterogeneity of the cell population and gene expression in complex biological systems.

This chapter primarily highlights the analytical problems in the evaluation of gene expression as a function of the cell cycle. For some approaches, it offers solutions. In general the term *gene expression*, as used here, means protein expression of intracellular and membrane-bound proteins.

II. METHODS

Analysis of gene expression at the protein level includes detection of intracellular or membrane-bound molecules. Detection of intracellular epitopes usually requires fixation of the cells, including the necessary membrane permeabilization. Typical procedures include alcohol fixation techniques (simultaneous fixation and permeabilization), or fixation with paraformaldehyde, followed by permeabilization with alcohol or detergent [7, 8]. All these techniques require controls indicating that the epitope is not disrupted by the fixation technique. In some cases, low concentrations of mild detergents might be sufficient for permeabilization of the cell membrane without fixation, thus leaving intracellular epitopes unaltered [9,10].

Demonstration of gene expression in viable cells is possible using the membrane-permeable, UV-excitable, DNA-fluorochrome Hoechst 33342 [11,12]. In a dual laser system, it is possible to analyze either surface receptor gene expression by using additional fluorochrome-labeled antibodies [13], or intracellular gene expression by applying membrane-permeable, fluorochrome-labeled, agonists [14]. Methods for Figs. 2 through 9 are described in detail elsewhere [15–21].

III. THE CELL CYCLE

Starting from fertilization, cells replicate their genome and distribute it equally to their progeny or daughter cells [22, 23]. Identical duplication of the nuclear genetic information is performed during the chromosomal cycle [24], consisting of the S, G_2, and M phases (Fig. 1). After mitotic division, cells enter the G_1 phase. At a putative transition point in G_1 phase the cells control their homeostatic, intracellular requirements and environmental signal cascades (see Fig. 1, G1B). After a positive decision, the cells enter the S phase, and so on. The most variable time of the cell cycle is the G_1 phase before the transition point (see Fig. 1, G1B), whereas the G_1 phase after this G1B-transition point, the S, G_2, and M phase durations are constant [25]. According to their ontogenetic requirements, cells may enter a G_0 phase after mitotic division. This could either be due to terminal differentiation, during which cells lose their replicative capabilities, or to a transient resting or dormant stage that the cells can leave after an equivalent stimulus [23].

The extracellular environment controls the cellular and replicative capabilities by interactions of different agonists (growth factors) or antagonists (inhibitors, chalones) with their corresponding receptors [26]. The intracellular-controlling elements consist of a complex network of kinases, phosphatases,

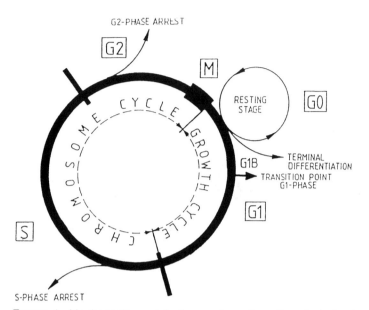

FIGURE 1 Mechanistic model of an eukaryotic cell cycle indicating the growth (G_1) and chromosome cycle $(S, G_2/M)$. The transition point (point of no return) is indicated in the G_1 phase by G_1B.

transcription factors, and last, but not least, proliferation-controlling elements, such as cyclins and cyclin-dependent kinases [27].

It is evident that most of the DNA replication-dependent enzymes are expressed in a cell cycle-correlated fashion [14, 28–31] compared with the mostly constitutive expression of membrane-bound, extracelluar receptors. On the contrary, cell cycle-correlated expression of genes as controlling elements is not a prerequisite for a regular signal cascade. The latter is primarily achieved by the enzymatic reactions of phosphorylation and dephosphorylation of already existing proteins [32].

With the advent of flow cytometry in the 1970s, the cell cycle could be dissected and displayed in its various compartments of the G_0/G_1, S, and G_2/M phases [4–6]. The application of special techniques allows the differentiation of the compartments G_0 and G_1 [33], and G_2 and M [34, 35]. In addition, incorporation of artificial nucleotides into the DNA of replicating cells enables analysis of actively proliferating versus arrested cells in the various cell cycle compartments [21, 36, 37].

Theoretically, gene expression might be quite complex and quantitatively differently regulated in all compartments or subcompartments of the cell cycle [38]. However, by applying various flow cytometric techniques, the cell cycle-related expression of biomolecules is easily analyzed using a variety of fixation and labeling methods [7, 8, 39]. Depending on the conditions of the experiment, detection of biomolecules in viable, as well as fixed, cells is possible using fluorochrome-labeled reporter molecules, for example, antibodies (cyclin), agonists (growth factors) or inhibitors (e.g., methotrexate).

IV. A SIMPLE VIEW OF CELL CYCLE-RELATED GENE EXPRESSION

Alteration of gene expression is common in cells undergoing ontogenetic development. The most typical example is lifelong in vivo hemopoiesis. Pluripotent bone marrow stem cells become more and more committed during subsequent replication cycles and start expression of new genes relevant for their physiological effector functions and immunological competence. The latter is evident in new or increased expression of surface markers [40]. This process can be artificially induced in vitro by cultivating human bone marrow cells in the presence of a cytokine cocktail.

An example of the alteration of gene expression by induction of cell differentiation and expression of new receptors on the cell surface is shown in Figure 2. The human bone marrow cells were labeled with CD14–FITC antibodies, fixed and permeabilized, and stained with the DNA fluorochrome propidium iodide (PI) (see Fig. 2 A and B). Immediately after setup of the cell culture, the human bone marrow culture contained no monocytes (see Fig. 2A), and only a few proliferating S-phase cells derived from the negative fraction. After exhaustion of the replicative potential (3 weeks in culture) 14.0% of differentiated CD14-positive monocytes are found. However, as expected, these cells are exclusively nonproliferating and are located in the G_0 compartment (see Fig. 2B). The few remaining cycling cells are derived from the undifferentiated CD14-negative cell population.

Receptor quantitation assays using different latex beads covered with different amounts of mouse antibodies were recently introduced [41]. With these standard beads and the correct controls, the quantification of the receptor numbers of cells has become feasible. Quantitative gene expression at the single-cell level can now also be achieved as a function of the cell cycle status of cells.

The simple cell cycle-related detection of intracellular biomolecules as a measure of gene expression is illustrated in Figure 3. This represents two typical examples of the analysis of replication-associated proteins: analysis of the Ki67 proliferation-associated antigen by a Ki67 antibody (see Fig. 3 A–C), and detection of p34^{cdc2} using a cdc2 antibody (see Fig. 3 D–F). As for the example in Figure 2, cells were fixed and permeabilized, and protein detection was performed by an indirect antibody-labeling technique [8,17]. Control cells were labeled with the secondary antibody only (see Fig. 3 A, D). In this experiment,

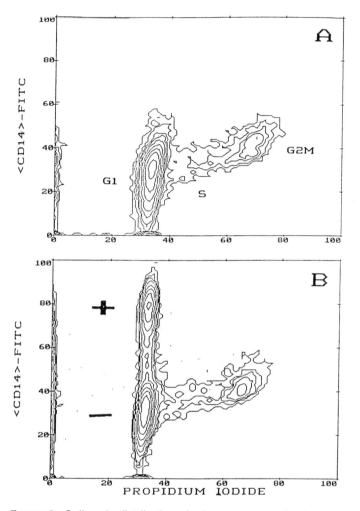

FIGURE 2 Cell cycle distribution of a human suspension bone marrow culture (A) in the inital stages and (B) after a 3-week period of proliferation in the presence of a cytokine cocktail. The induction of gene expression is shown by the differentiation of proliferating progenitor cells into noncycling CD14-positive monocytes. X-axis, linear; Y-axis, log.

human peripheral blood lymphocytes were activated by polyclonal stimulation with phytohemagglutinin (PHA), and analysed 48 and 72 h later. After 48 h the intracellular amount of both the Ki67 antigen, as well as the p34^{cdc2} protein increased significantly in cycling G_1, S, and G_2/M cells (see Fig. 3 B, E). However, a significant number of cells still remained in the negative, nonactivated,

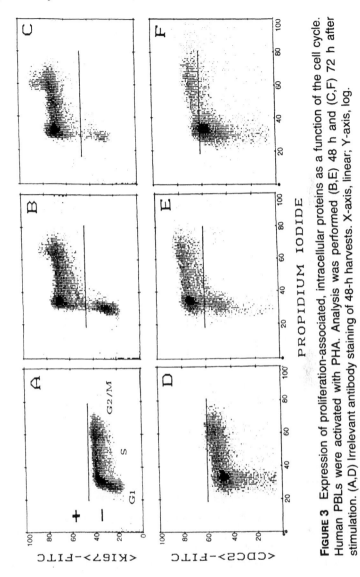

FIGURE 3 Expression of proliferation-associated, intracellular proteins as a function of the cell cycle. Human PBLs were activated with PHA. Analysis was performed (B,E) 48 h and (C,F) 72 h after stimulation. (A,D) Irrelevant antibody staining of 48-h harvests. X-axis, linear; Y-axis, log.

noncycling fraction of G0 cells: 24% for the Ki67 antigen and 14% for the p34^{cdc2} protein. The constitutive, asynchronous and, therefore, late activation of lymphocytes is obvious in Figure 3C and F. The numbers of negative, nonproliferating cells in the 72-h harvest decreased to 8% for Ki67 and 6% for p34^{cdc2}. By comparing the expression of the Ki67 antigen in cycling cells between the 48- and 72-h harvest, it is evident that the intracellular amount of this antigen remains constant (see Fig. 3 B,C). On the contrary, the expression of p34^{cdc2} seems to decrease slightly in cycling lymphocytes after 72 h, as shown by the lower cdc2 fluorescence labeling (see Fig. 3 E,F).

V. THE PROBLEM OF CELLULAR AMOUNT AND CONCENTRATION

As shown by the two foregoing examples, conventional flow cytometric techniques enable fast and simple analysis of cell cycle-related gene expression in cell populations. However, this kind of analysis ignores the problem of altered cell size (e.g., during the activation of lymphocytes in Fig. 3). The cell size increases significantly during activation of small, resting cells that become large, blast cells. However, even during the cell cycle, cellular biomass and, therefore, cell size, increases continuously from G_1, to S, to G_2/M phases [42, 43]. Correct cell size analysis can be performed with Coulter volume analysis. Only a few flow cytometers offer this analytical option. However, another parameter, forward light scatter (FSC), although not strictly linearly related to cell size, also gives a good qualitative measure of cell size and changes thereof [44]. In this chapter and the subsequent Figures 4–7, FSC will be used as a cell size-related parameter.

In Figure 4 the increase of monochlorobimane fluorescence, as a measure of thiols of glutathione (GSH) and SH-proteins, is shown as a function of the cell cycle of PHA-activated lymphocytes. It is evident from Figure 4A that the resting, nonactivated small lymphocytes display a low GSH/protein content (G_0 phase). Activated lymphocytes in G_1 phase exhibit an increased fluorescence, as do the cycling G_1-phase cells. During S and G_2/M phases the GSH/protein content increases continuously. This kind of analysis represents a quantitation of the amount of biomolecules in individual cells.

The cell cycle-related expression pattern of the cellular GSH/protein content changes if one takes into account the alteration of cell size. In Figure 4B the GSH–protein-related monochlorobimane fluorescence was divided by the FSC signal in real-time in the data acquisition computer and displayed as the Y-axis signal. The latter now represents a concentration value of the analyzed biomolecules, instead of a cellular amount. If one compares Figures 4A and B it becomes clear that both the amount and the concentration increase during lymphocyte activation (G_0 to G_1). Therefore, the increase of gene expression analyzed as the cellular amount of GSH–protein is not simply due to the

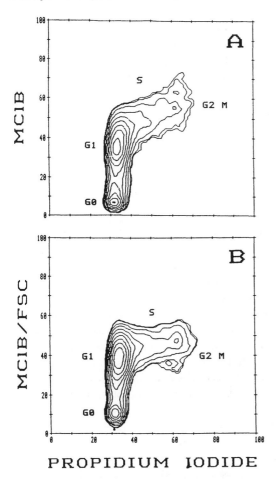

PROPIDIUM IODIDE

FIGURE 4 Cell activation and proliferation-dependent increase of gene expression related to the increase in cell size. Human PBLs were activated with PHA. Synthesis of glutathione and thiolproteins was analyzed after 72 h using monochlorobimane (MCIB) staining. (A) Increase of intracellular amount (MCIB-fluorescence). (B) Increase of intracellular concentration [MCIB fluorescence divided by FSC (cell size)]. X-axis, linear; Y-axis, log.

phenomenon of an increasing cell size. In addition this indicates that both a critical threshold of the intracellular amount as well as the concentration of GSH-proteins are necessary for entry into G_1 toward S phase.

The major differences are seen in S and G_2/M phases. The increase of the concentration is much smaller compared with the increase of the amount (see Fig. 4B vs. A). From this data, it can be concluded that the increase of GSH-protein during S and G_2/M phases is related to the changes in the cell size only. There seems to be no significant net synthesis over the increase of the total biomass compared with the changes during the G_0–G_1 transition.

Quantitation of the intracellular amount of biomolecules and division by the FSC signal gives information about a concentration value. Similarly, the calculation of the amount of fluorescence of extracellular, membrane-bound molecules divided by FSC results in a density value. One such example, also displaying a significant difference in cell cycle-related gene expression, as represented by either the amount or the density of a biomolecule, is shown in Figure 5. In this example, the surface IgG expression in viable, exponentially growing mouse hybridomas was analyzed, using the membrane-permeable fluorochrome Hoechst 33342. The increase in the amount of membrane-bound IgG obviously parallels the increase in cell size, and no abrupt difference of expression is seen between cells in different cell cycle compartments (see Fig. 5A). However, calculating the IgG fluorescence/FSC values (see Fig. 5B), surprisingly reveals that the density of the IgG expression on the cell surface decreases continuously as cells move from G_1 into S and G_2/M phases. Therefore, the net increase of the IgG expression during the cell cycle, as evident from Figure 5A, is smaller compared with the parallel increase of the cell size of hybridomas. Numerous studies indicate cell cycle-related IgG expression or synthesis in mouse hybridomas [44–47]. However, the cell size-related parameter still needs to be evaluated in a direct flow cytometric assay.

A third example of alteration of gene expression analysis related to cell size is demonstrated by dihydrofolate reductase (DHFR) expression in Chinese

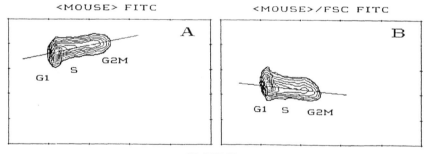

H 33342

FIGURE 5 Expression of membrane-bound immunoglobulins of mouse hybridomas as a function of the cell cycle and cell size. Exponentially growing mouse hybridomas were labeled with antimouse F(ab)$_2$ IgG–FITC antibodies, and the the cell cycle of the viable cells was recorded using Hoechst 33342. (A) Amount of membrane-bound IgG (IgG–FITC); (B) density-concentration of membrane-bound IgG [IgG–FITC divided by FSC (cell size)]. *X*-axis, linear; *Y*-axis, log.

hamster ovary (CHO) cells. A DHFR-negative CHO cell line was transfected with a vector carrying the *dhfr* gene. This gene was amplified through selection pressure by applying increasing medium concentrations of the DHFR inhibitor methotrexate (MTX) [48]. The intracellular level of DHFR expression in viable cells was analyzed using the membrane-permeable, fluorochrome-labeled MTX–FITC [14, 49], and the cell cycle was displayed by the DNA fluorochrome, Hoechst 33342.

Figure 6A demonstrates the conventional cell cycle-related expression of DHFR. In the first instance, as shown in Figure 6A, the distribution of the MTX–FITC fluorescence and, therefore, the amount of intracellular DHFR, is skewed toward higher values and is unimodal. Only cells exhibiting increased amounts of DHFR move into S and, subsequently, into G_2/M phases. The rational behind this is simple: only cells with increased amounts of a key enzyme for thymidine nucleotide synthesis start the S phase. The DHFR gene expression, as a function of cell size and cell cycle in Figure 6B reveals a different pattern. The MTX–FITC (DHFR) distribution in G_1-phase cells is even more skewed toward higher concentrations. In addition, the cycling S-phase cells do not have the highest concentration of DHFR. Although the concentration of DHFR in cells entering S phase is slightly increased over the mean of the DHFR distribution in G_1-phase cells, it is obvious that it is significantly lower compared with the maximum DHFR concentration.

Therefore, the CHO cells seem to regulate entry into S phase by the amount of DHFR, and not by its concentration. This is different from activated peripheral blood lymphocytes (PBLs), which start S phase only at the highest GSH–protein content as well as concentration (see Fig. 4). The deviation of cell cycle- and cell size-regulated gene expression in CHO cells as prerequisite for entry into S phase is revealed in Figure 7. In Figure 7A, gates were set on the cell fractions exhibiting the highest (gate 7), moderate (gate 6), and lowest (gate 5) DHFR concentrations. The corresponding forward scatter values (FSC, cell size) were recorded and are displayed in Figure 7B. The cells with the highest DHFR concentrations exhibit the smallest FSC values (which correspond to the smallest cell size). The fractions with moderate and low DHFR concentrations are the intermediate and largest cell sizes, respectively. It may be speculated that the differences on the large-sized cells arise from unequal divisons during mitosis. The smallest cell fraction (gate 7) needs to grow to a critical threshold cell size before S-phase entry [42], in addition to the increase of DHFR up to its critical cytoplasmic amount.

VI. THE PROBLEM OF TWO INDEPENDENT KINETIC PROCESSES

An additional problem arises in the interpretation of cell cycle-related gene and protein expression that is induced by exogenous activators. In principle, there is

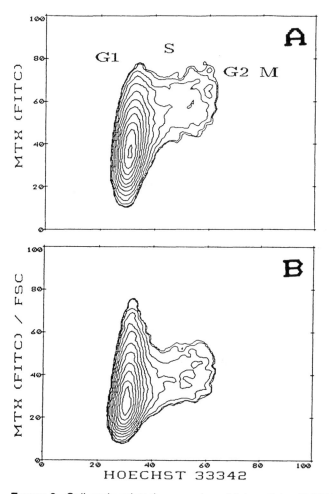

FIGURE 6 Cell cycle-related expression of intracellular DHFR in viable exponentially growing CHO cells as a function of cell size. The cells were stained with membrane-permeable MTX–FITC (DHFR) and Hoechst 33342 (DNA). (A) Amount of intracellular DHFR (MTX–FITC); (B) concentration of intracellular DHFR [MTX–FITC divided by FSC (cell size)]. X-axis, linear; Y-axis, linear.

a delay between the induction process and analysis. During this time two kinetic processes run in parallel: the cells progress further through the cell cycle compartments, and there is a concomitant increase in the amount of a synthesized protein. How can it be shown at what time and at what cell cycle stage the gene-protein expression started, and what would be its extent?

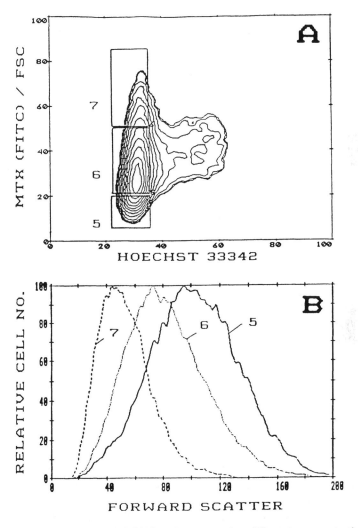

FIGURE 7 Cell size of CHO cells expressing different concentrations of DHFR. G_1-phase cell populations exhibiting different intracellular DHFR concentrations were gated in regions 5, 6, and 7, and the corresponding cell sizes were recorded. (A) Concentration of intracellular DHFR (MTX–FITC divided by FSC); (B) FSC distribution (cell size) of the CHO populations gated in panel A. X-axis, linear; Y-axis, linear.

This problem is illustrated in Figure 8. MPC-11 mouse plasmacytoma cells, negative for the intermediate filament protein vimentin, can be induced by the phorbol ester 12-*O*-tetradecanoylphorbol-13-acetate (TPA) to start vimentin synthesis only 2 h after induction [16]. In Figure 8A the cell cycle distribution of the exponentially growing cell population at time zero (addition of phorbol ester) is displayed. Although, in this control culture, there is no vimentin expression visible, mRNA and vimentin synthesis starts soon after (kinetic process t_2). However, in addition to the translation of the vimentin mRNA, the cells continuously move through the cell cycle (kinetic process t_1). They become arrested in the next G_1 phase after mitosis [1, 16, 20].

Analysis of the 8-h harvest of the TPA-treated culture is displayed in Figure 8B. It is obvious that significant fractions of the cells in all of the cell cycle compartments exhibit intracellular vimentin expression. This is especially evident for the G_1-phase fraction. However, where were these cells at the time of TPA treatment 8 h ago? Were they in early, mid, late S, or G_2, or could this vimentin-positive G_1 fraction even be composed of arrested 0-h G_1 and cycling S and G_2 cells? To reveal at least the proliferative complexity the DNA of the cells must be continuously labeled with bromodeoxyuridine (BRDU) to trace their initial cycling status and replicative history [21]. The BRDU/Hoechst–PI quenching technique is performed under mild conditions, allowing immunocytochemical staining of various membrane-bound and intracellular epitopes [20, 21, 50].

A typical experiment to solve this problem of two independent kinetic processes is shown for the TPA-treated plasmacytoma cells in Figure 9. The plasmacytoma cells were cultured for 10 h as control culture (see Fig. 9A, C) or treated with TPA to induce vimentin synthesis. It is evident from Figure 9A that almost all cells have left the first cell cycle, except a few G_2/M cells. The cells in the second cycle G_1' phase are derived from the first cycle G_1 phase. All the cells that started from the first cell cycle S phase into the second cycle G_1' phase have already moved into S' and G_2/M' phases. Plasmacytoma cells treated with TPA become arrested in the second cycle G_1' phase (see Fig. 9B). The most advanced cells in the first cell cycle (S and G_2/M) have already divided and are represented by the G_1'–G_1 lane. The G_1-phase cells in the first cell cycle before the transition point also become arrested [1]. Cells after this transition point (G_1B, see Fig. 1) progressed into the first cell cycle S and G_2/M phases, and are found in the left peak of the second cycle G_1'-phase lane. It is evident that this type of BRDU/Hoechst-PI quenching technique clearly reveals the replicative history [20, 21, 37, 50].

The gene expression status of these cells is displayed in Figure 9C and D. In the control culture, almost all cells are vimentin-negative (see Fig. 9C). The left arrow represents cells in the G_1' phase, and the right arrow cells in the G_2/M' phase of the second cycle. The cells in between represent second-cycle S'-phase

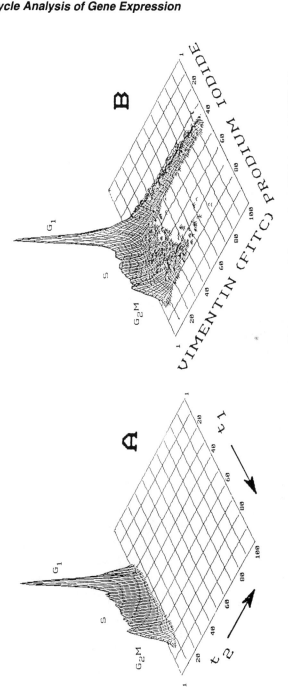

FIGURE 8 Increase of cell cycle-dependent expression of vimentin in exponentially growing, TPA-treated MPC-11 mouse plasmacytoma cells. (A) Untreated, vimentin-negative, control culture at setup of the experiment. The X- and Y-axis indicate the simultaneously existing kinetics of proliferation (t_1) and gene expression (t_2), respectively. (B) Culture treated for 8 h. X-axis, linear; Y-axis, linear.

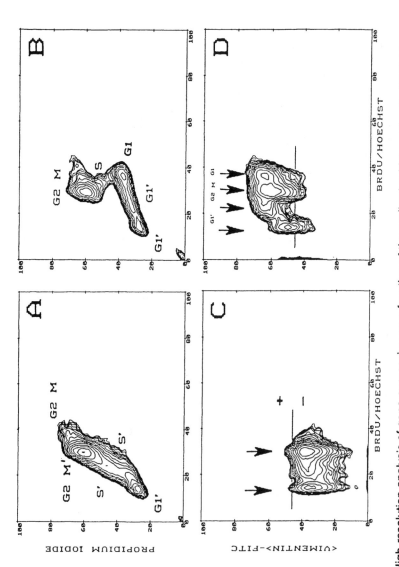

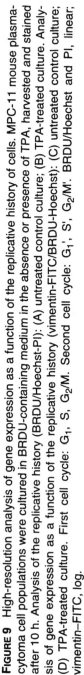

FIGURE 9 High-resolution analysis of gene expression as a function of the replicative history of cells. MPC-11 mouse plasmacytoma cell populations were cultured in BRDU-containing medium in the absence or presence of TPA, harvested and stained after 10 h. Analysis of the replicative history (BRDU/Hoechst-PI): (A) untreated control culture; (B) TPA-treated culture. Analysis of gene expression as a function of the replicative history (vimentin-FITC/BRDU-Hoechst): (C) untreated control culture; (D) TPA-treated culture. First cell cycle: G$_1$, S, G$_2$/M. Second cell cycle: G$_1$', S', G$_2$/M'. BRDU/Hoechst and PI, linear; vimentin–FITC, log.

cells. After 10-h treatment with TPA most plasmacytoma cells express vimentin in all compartments of the cell cycle (see Fig. 9D). However, the expression rates are different. The lowest expression is found in cells derived from G_1-phase cells at culture setup (indicated by the far left arrow). Cells arrested in the first cycle G_1 phase before the transition point G_1B (far right arrow) express higher levels, as do the cycling cells in the G_2/M phase (near right arrow). The near left arrow indicates cells in the second cycle G_1' phase, which were in the first cell cycle mid-S phase at TPA treatment. Although there is a 10-h delay between TPA treatment and cell analysis, it is evident from the horseshoe-shaped pattern in Figure 9D that the vimentin gene induced by TPA is higher in the S-phase cells than in the G_1-phase cells in the first cell cycle [20].

As shown in Figure 9, gene expression rates may be different, but still substantial, in all cell cycle compartments. If the analysis is significantly delayed after the start of treatment, an equalized-balanced amount of induced gene product may mask the initial difference at later time points. The solution to this problem requires either early analysis or stathmokinetic techniques [e.g., Colcemid or thymidine arrest of cells] in addition to the flow cytometric analysis of the replicative history.

VII. CONCLUSIONS

Gene expression as indicated by protein expression may be constitutive during the cell cycle (e.g., housekeeping proteins of the energy metabolism), or regulated differently in various cell cycle compartments (e.g., proteins relevant for DNA synthesis). However, although not shown in the foregoing examples and not yet detectable by flow cytometry, neither the constitutive presence of proteins, nor their distinct expression in different cell cycle compartments gives final evidence for their functional activity: protein functions might be regulated by phosphorylation, dephosphorylation, associations of subunits, or other modifications [26, 32].

In principle, the critical task remains whether the cellular amount of an expressed gene, its cellular concentration, or even its local density is decisive for regulatory processes in cells. As shown earlier, a solution to these questions might be the correlation of the amount with the cell size by FSC but if possible, the more size-related parameter "Coulter volume" should be used. In most cases the alteration of the amount and concentration of a protein runs in parallel. Deviations, however, might be observed if unequal cytokinesis occurs after mitosis.

The alteration of the amount of secreted proteins often parallels the increase in cell size. However, the fact that the density-concentration of membrane-bound immunoglobulins decreases as cells move through the cell cycle (see Fig. 5) is not easy to interpret. On the one hand, the cell size of hybridomas might increase more rapidly than the amount of membrane-bound IgGs: on the other hand, the secretion rate might speed up as cells move from G_1 through S and G_2/M phases. Unfortunately, for physiological processes such as secretion of proteins, no stathmokinetic techniques are available, as they are for cell proliferation.

In model systems in which gene expression is induced in cycling cells by an activator, two, possibly independent, kinetic processes occur: progression through the cell cycle and an increase in the amount of the expressed protein (see Fig. 8). The time delay between induction and analysis often masks cell cycle-related differences in gene expression. This problem may be solved only if one traces the replicative history of cells by a nucleic acid-labeling technique (BRDU labeling; see Fig. 9) in parallel with protein expression. For constitutive, although quantitatively different, protein synthesis in different cell cycle compartments, this type of analysis must be performed in short time intervals or by the use of stathmokinetic techniques.

In general, alteration of gene expression is accompanied by changes in a variety of other cell physiological parameters which, in addition to kinetic aspects, may be of relevance for correct analysis and interpretation as a function of the cell cycle progression. Awareness of this complexity makes correct interpretation of cell cycle-related gene expression even more significant.

REFERENCES

1. G. Giese, M. Kubbies, and P. Traub, Cell cycle dependent vimentin expression in elutriator-synchronized, TPA-treated MPC-11 mouse plasmacytoma cells, *Exp. Cell Res. 200*:118 (1992).
2. Z. Darzynkiewicz, F. Traganos, and M. Kimmel, Assay of cell cycle kinetics by multivariate flow cytometry using the principle of stathmokinesis, *Techniques in Cell Cycle Analysis* (J.W. Gray and Z. Darzynkiewicz, eds.), Humana Press, Clifton, NJ, 1987, p. 291.
3. J. K. Heath, ed., *Growth Factors*, IRL Press, Oxford University Press, Oxford, 1993.
4. M. R. Melamed, T. Lindmo, and M. L. Mendelsohn, eds., *Flow Cytometry and Sorting*, 2nd ed., Wiley-Liss, New York, 1990.
5. K. D. Bauer, R. E. Duque, and T. V. Shankey, eds., *Clinical Flow Cytometry: Principles and Application*, Williams & Wilkins, Baltimore, 1993.
6. H. M. Shapiro, ed., *Practical Flow Cytometry*, 3rd ed., Wiley-Liss, New York, 1994.
7. J. P. Robinson, ed., *Handbook of Flow Cytometry Methods*, Wiley-Liss, New York, 1993.
8. M. Kubbies, Flow cytometry, *Methods of Immunological Analysis* (R. F. Masseyeff, W. H. Albert, and N. A. Staines, eds.), VCH Verlag Weinheim, 1993, p. 376.
9. M. Assenmacher, Combined intracellular and surface staining, *Flow Cytometry and Cell Sorting* (A. Radbruch, ed.), Springer Verlag, Berlin, 1992, p. 53.
10. M. R. Miller, J. J. Castellot, and A. B. Pardee, A general method for permeabilizing monolayer and suspension cultured animal cells, *Exp. Cell Res. 120*:421 (1979).

11. D. J. Arndt-Jovin and T. M. Jovin, Analysis and sorting of living cells according to deoxyribonucleic acid content, *J. Histochem. Cytochem. 25*:585 (1977).

12. M. J. Lydon, K. D. Keeler, and D. B. Thomas, Vital DNA staining and cell sorting by flow microfluorometry, *J. Cell. Physiol. 102*:175 (1980).

13. M. R. Loken, Simultaneous quantitation of Hoechst 33342 and immunofluorescence on viable cells using a fluorescence activated cell sorter, *Cytometry 1*:136 (1980).

14. M. Kubbies, H. Stockinger, Cell cycle dependent DHFR and TPA production in cotransfected, MTX-amplified CHO cells revealed by dual laser flow cytometry, *Exp. Cell Res. 188*:267 (1990).

15. D. R. Van Bockstaele, F. Lardon, H. W. Snoeck, M. E. Peetermans, and M. Kubbies, BrdU-Hoechst ethidium bromide quenching technique for studying kinetics of hematopoiesis, *Blood 80*:289 (1992).

16. G. Giese, M. Kubbies, and P. Traub, Alterations of cell cycle kinetics and vimentin expression in TPA-treated, asynchronous MPC-11 mouse plasmacytoma cells. *Exp. Cell Res. 190*:179 (1990).

17. H. Kreipe, H. J. Heidebrecht, S. Hansen, W. Roehlk, M. Kubbies, H. H. Wacker, M. Tiemann, H. J. Radzun, and M. Parwaresch, A new proliferation-associated nuclear antigen detectable in paraffin embedded tissues by the monoclonal antibody Ki-S1, *Am. J. Pathol. 142*:3 (1993).

18. G. C. Rice, E. A. Bump, D. C. Shrieve, W. Lee, and M. Kovacs, Quantitative analysis of cellular glutathione by flow cytometry utilizing monochlorobimane: Some applications to radiation and drug resistance in vitro and in vivo, *Cancer Res. 46*:6105 (1986).

19. M. Kubbies, Flow cytometric recognition of clastogen induced chromatin damage in G0/G1 lymphocytes by non-stoichiometric Hoechst fluorochrome binding, *Cytometry 11*:386 (1990).

20. G. Giese, M. Kubbies, and P. Traub, High resolution cell cycle analysis of cell cycle correlated vimentin expression in asynchronously grown, TPA-treated MPC-11 cells by the novel flow cytometric multiparameter BrdU-Hoechst/PI and immunolabeling technique, *J. Cell. Physiol. 161*:209 (1994).

21. M. Kubbies, High-resolution cell cycle analysis: The flow cytometric BrdU-Hoechst quenching technique, *Flow Cytometry and Cell Sorting* (A. Radbruch, ed.), Springer Verlag, Berlin, 1992, p. 75.

22. D. M. Prescott, *Reproduction of Eucaryotic Cells*, Academic Press, New York, 1976.

23. R. Baserga, *The Biology of Cell Reproduction*, Harvard University Press, Cambridge, MA, 1985.

24. J. M. Mitchinson, ed., *The Biology of the Cell Cycle*, Cambridge University Press, Cambridge, 1971.

25. J. A. Smith and L. Martin, Do cells cycle? *Proc. Natl. Acad. Sci. USA 70*:1263 (1973).

26. R. A. Bradshaw and S. Prentis, eds., *Oncogenes and Growth Factors*, Elsevier, Amsterdam, 1987.

27. J. Pines, The cell cycle kinases, *Semin. Cancer Biol. 5*:305 (1994).

28. D. T. Denhardt, D. R. Edwards, and C. L. J. Parfett, Gene expression during the mammalian cell cycle, *Biochim. Biophys. Acta 865*:83 (1986).

29. T. Stokke, H. Holte, B. Erikstein, C. L. Davies, S. Funderud, and H. B. Steen, Simultaneous assessment of chromatin structure, DNA-content and antigen expression by dual wavelength excitation flow cytometry, *Cytometry 12*:172 (1991).

30. G. Landberg, E. M. Tan, and G. Ross, Flow cytometric multiparameter analysis of proliferating cell nuclear antigen/cyclin and Ki-67 antigen: A new view of the cell cycle, *Exp. Cell Res. 187*:111 (1990).

31. J. Gong, F. Traganos, and Z. Darzynkiewicz, Expression of cyclins B and E in individual MOLT-1 cells and in stimulated human lymphocytes during their progression through the cell cycle, *Int. J. Oncol. 3*:1037 (1993).

32. T. Boulikas, The phosphorylation connection to cancer, *Int. J. Oncol. 6*:271 (1995).

33. Z. Darzynkiewicz, F. Traganos, T. K. Sharpless, and M. R. Melamed, Cell cycle related changes in nuclear chromatin of stimulated lymphocytes as measured by flow cytometry, *Cancer Res. 37*:4635 (1977).

34. J. K. Larson, B. Munch-Petersen, J. Christiansen, and K. Jorgensen, Flow cytometric discrimination of mitotic cells: Resolution of M, as well as G_1, S and G_2 phase nuclei with mithramycin, propidium iodide and ethidium bromide after fixation with formaldehyde, *Cytometry 7*:54 (1986).

35. H. Koch, T. Bettecken, M. Kubbies, D. Salk, J. W. Smith, and P. S. Rabinovitch, Flow cytometric analysis of small DNA content differences in heterogeneous cell populations: Human amniotic fluid cells, *Cytometry 5*:118 (1984).

36. J. W. Gray and B. H. Mayall, eds., *Monoclonal Antibodies Against Bromodeoxyuridine*, Alan R. Liss, New York, 1985.

37. M. G. Omerod and M. Kubbies, Cell cycle analysis of asynchronous cell populations by flow cytometry using bromodeoxyuridine label and Hoechst–propidium iodide stain, *Cytometry 13*:678 (1992).

38. J. W. Jacobberger, Cell cycle expression of nuclear proteins, *Flow Cytometry: Advanced Research and Clinical Applications*, Vol. 1 (A. Yen, ed.), CRC Press, Boca Raton, FL, 1989, p. 305.

39. Z. Darzynkiewicz, J. P. Robinson, and H. A. Crissman, eds., *Methods in Cell Biology: Flow Cytometry*, 2nd ed., part A, Academic Press, San Diego, 1994.

40. C. I. Civin and S. D. Gore, Antigenic analysis of hematopoiesis: A review, *J. Hematother. 2*:137 (1993).

41. R. A. Hoffmann, D. J. Recktenwald, and R. F. Vogt, Cell-associated receptor quantification, *Clinical Flow Cytometry: Principles and Application* (K. D. Bauer, R. E. Duque, and T. V Shankey, eds.), Williams & Wilkins, Baltimore, 1993.

42. R. Shields, R. F. Brooks, P. N. Riddle, D. F. Capellaro, and D. Delia, Cell size, cell cycle and transition probability in mouse fibroblasts, *Cell 15*:469 (1978).

43. M. Al-Rubai, S. Chalder, R. Bird, and A. N. Emery, Cell cycle, cell size and mitochondrial activity of hybridoma cells during batch cultivation, *Cytotechnology 7*:179 (1991).

44. W. M. Grogan and J. M. Collins, eds., *Guide to Flow Cytometry Methods*, Marcel Dekker, New York, 1990, p. 86.

45. E. Meilhoc, K. D. Wittrup, and J. E. Bailey, Application of flow cytometric measurement of surface IgG in kinetic analysis of monoclonal antibody synthesis and secretion by murine hybridoma cells, *J. Immunol. Methods 121*:1167 (1989).

46. M. Al-Rubai and A. N. Emery, Mechanisms and kinetics of monoclonal antibody synthesis and secretion in synchronous and asynchronous hybridoma cell cultures, *J. Biotechnol. 16*:67 (1990).

47. S. J. Kromenaker and F. Scrienc, Cell cycle dependent protein accumulation by producer and non-producer murine hybridoma cell lines: A population analysis, *Biotechnol. Bioeng. 38*:665 (1991).

48. U. H. Weidle, P. Buckel, and J. Wienberg, Amplified expression constructs for human tissue-type plasminogen activator in CHO cells: Instability in the absence of selective pressure, *Gene 66*:193 (1988).

49. G. B. Henderson, A. Russell, and J. M. Whiteley, A fluorescent derivative of methotrexate as an intracellular marker for dihydrofolate reductase in L1210 cells, *Arch. Biochem. Biophys. 202*:29 (1980).

50. P. S. Rabinovitch, M. Kubbies, Y. C. Chen, D. Schindler, and H. Hoehn, BrdU-Hoechst flow cytometry: A unique tool for quantitative cell cycle analysis, *Exp. Cell Res. 174*:309 (1988).

5

Stability of Monoclonal Antibody Production in Hybridoma Cell Culture

JOSÉ M. COCO MARTIN
The Netherlands Cancer Institute, Amsterdam,
The Netherlands

E. COEN BEUVERY
National Institute for Public Health
and Environmental Protection (RIVM),
Bilthoven, The Netherlands

I. INTRODUCTION

Standardization and control of pharmaceutical proteins during their production is of major interest for producers and regulatory authorities [1]. Important aspects of the production of biologicals by mammalian cells for pharmaceutical applications in humans are the consistency and biological activity of the produced proteins, and the stability of the product formation and cell line during long-term culture. For this reason, these proteins should be characterized at several stages during the development and production processes of the proteins, including the establishment, standardization, and selection of the cells; adaption to culture medium; and upscaling to larger-sized culture systems. Moreover, the stability of the cells and the product during long-term culturing, for at least several months, and the product stability during downstream processing and after long term-storage have to be assessed.

To obtain information about the stability of both the produced protein and the cell line, a panel of appropriate analytical methods is required. These analytical methods may include flow cytometric (FC) [2–4], biochemical [sodium dodecyl sulfate–polyacrylamide gel electrophoresis (SDS–PAGE), isoelectric focusing (IEF); 5,6], or immunochemical methods [enzyme-linked immunosorbent assay (ELISA), sol particle immuno assay (SPIA); 7–9]. The combination of these different analytical methods enables us to determine not only the stability of the cell, but also that of the product formation.

The introduction of FC has facilitated the use of fluorochrome-conjugated antibodies for the characterization of both mammalian cells [10,11] and bacteria. The FC measurement allows rapid analysis of cell populations, and large numbers of cell samples can be quantitatively analyzed.

Initially, FC has had a significant impact on the study of cell populations of the immune system, such as B cells [3] or T lymphocytes [2,12] and in cancer research. Flow cytometry is not only an excellent tool to determine relative amounts of cell surface antigens [13], but it is also suitable for the determination of cytoplasmic antigens [14] and nuclear components [15,16]. Jacobberger et al. [4] described the quantification of the SV40 T-antigen in the cytoplasm of SV40-transformed and lytically infected cells.

During the last decade FC has entered the field of biotechnology (e.g., for selection of isotype switch variants of hybridoma cells [17] or quadromas [18] and as analytical tool during large-scale cultivation of mammalian cells used for the production of biologicals [19–22]).

This chapter will focus on the use of FC to characterize the stability of monoclonal antibody (MAb) production in batch as well as continuous culture systems.

II. MATERIALS AND METHODS

A. Methanol Fixation of Cell Suspensions (Cytoplasmic IgG Content)

Cell suspensions (10^6/ml) were washed three times with cold (4°C) phosphate-buffered saline (PBS), pH 7.3. After the final washing and resuspension, the cells were fixed and permeabilized by adding 1 ml methanol (Merck, Darmstadt) at −70°C while gently stirring. Permeabilization and fixation were performed with 100% methanol. According to Levitt et al. [26], fixation with 100% methanol is excellent for preserving both cell morphology and fluorescent staining. Fixed cells were kept at −20°C for at least 15 h, and were washed three times with PBS, pH 7.3, containing 0.5% bovine serum albumin (BSA; PBS–BSA). The cells were incubated with fluorescein isothiocyanate (FITC)-labeled goat antimouse IgG1, IgG2a, IgG2b, IgG3, and light (κ) chain-specific antibodies (Southern Biotechnology Association, AL), or with a synthetic peptide–fluorescein conjugate, for 1 h at 37°C. After the final washes the cells were resuspended in 1 ml PBS–BSA.

B. Paraformaldehyde Fixation and Staining of Cell Suspensions (Membrane IgG Content)

Cell suspensions (10^6/ml) were washed twice with PBS–BSA, and incubated for 40 min at 0°C with FITC-labeled goat antimouse IgG2a chain-specific antibodies (Southern Biotechnology Association, AL), or with a synthetic peptide–fluorescein conjugate.

After two washes the cell suspensions were fixed with 0.25% paraformaldehyde (BDH, Dorset).

C. Flow Cytometric Analysis

Samples were analyzed with a flow cytometer (FACScan; Becton Dickinson BV, Etten-Leur, The Netherlands). An HP310 computer (Hewlett-Packard Corporation, Pittsburgh, PA) with a Consort 30 program (Becton Dickinson) was used for data processing. Dead cells, debris, and cell clumps were gated out, according to their forward–angle- and right–angle-scattering properties. Windows were set to obtain the percentages of each population present in samples.

D. Detection of Antibody-Secreting Cells

Wells of microtiter plates (Costar, Cambridge, MA) were coated for 3 h at 37°C with monoclonal rat antimouse κ-light chain-specific antibodies (30 μg/ml in PBS, pH 7.3). The plates were washed three times with sterile PBS, pH 7.3. Cell suspensions (750 cells per ml) were added (100 μl) to each well. After incubation for 20 h at 37°C under sterile conditions in a CO_2 incubator, the plates were washed thoroughly. Incubation for 1 h at 37°C was performed with biotinylated monoclonal rat antimouse IgG2a-specific antibodies (1:500; Sanbio BV, Uden, The Netherlands). This was followed by incubation with an alkaline phosphatase–avidin (Sigma, St Louis, MO) conjugate (100 μl ; 1:1000), also for 1 h at 37°C. After the final washings, 100 μl of 1-mg/ml enzyme substrate, 5-bromo-4-chloro-3-indolyl phosphate (Sigma, St Louis, MO) in 2-amino-2-methyl-1-propanol (Sigma) plus 0.6% agarose was added to the wells. The spots were either counted under a microscope, using 14 × magnification, or were analyzed with an image-processing system.

III. RESULTS AND DISCUSSION

A. General Aspects

Typical cytograms of cytoplasmic and membrane IgG content of antibody-producing hybridoma cells are shown in Figure 1. The population with the low fluorescence intensity represents the negative control (see Fig. 1A). Negative controls can be either only methanol-fixed cells or cells incubated with a nonspecific antibody. Nonproducing hybridoma or nonproducing myeloma cells (e.g., the fusion partner of the B cells) can also be used for this purpose (Fig. 2). Incubation with anti-heavy chain-specific antibodies as well as with anti–light (κ) chain-specific antibodies allows discrimination between the expression of the heavy and light chains within hybridoma cells. Figure 2 shows that the ratio of heavy/light chains is less than 1. This indicates that fewer heavy chains than light chains are synthesized by the cells. This moderate overexpression could be favorable for cell survival and viability. The population with the high mean fluorescence intensity is made up of cells that react with the anti–IgG-specific antibody and that reflect the intracellular or membrane IgG content (see Fig. 1B). These cells are able to synthesize IgG and probably to secrete the IgG into the culture fluid. The latter, however, can be detected by other immunochemical or biochemical methods (e.g., ELISA, SPIA, SDS–PAGE). The mean fluorescence intensity can be a measure of the expression of the IgG in the cytoplasm as well as on the membrane. The distribution of the fluorescence intensity reflects the heterogeneity of different levels of expression on the basis of the single cells present in the population (Fig. 3). The sharper the distribution, the less variation, indicating that a more homogeneous population is present in the culture. The broader the distribution, the more heterogeneous cells are present in the culture. Also, a bimodal distribution can be observed if the cell population consists of high, low, and non–antibody-synthesizing cells (see Fig. 3a). The cytoplasmic and membrane IgG content could be correlated with other methods that determine the antibody production and especially the IgG secretion at the single-cell level [23].

B. Batch Culture

The MN12 cell line is used for several of our studies on the stability of antibody formation and secretion. These studies were performed in batch as well as continuous culture systems. In a batch culture system, the cytoplasmic IgG content showed the same results as those for the antibody-secreting cells (Fig. 4), indicating that all the cells that form antibody secrete this into the culture fluid. This makes FC a rapid method for the determination of antibody-synthesizing, as well as antibody-secreting, cells. However, discrepancies between cytoplasmic-positive cells and antibody secretion are also observed [24]. This makes it necessary to characterize the cell lines with other immunochemical

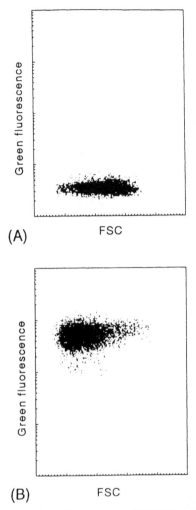

FIGURE 1 Cytograms of MN12 hybridoma cells (A) after methanol fixation and (B) after methanol fixation followed by incubation with an anti-IgG2a-specific antibody.

methods. The relative cytoplasmic IgG content and specific activity of cytoplasmic IgG, studies in a batch culture, were determined after incubation with fluorescein-labeled goat antimouse IgG2a antibodies or a peptide–fluorescein conjugate, respectively. The peptide consists of a sequence of amino acids that is recognized by the MN12 MAb; therefore, it can be used to determine the biological activity of this antibody [25]. Figure 5A shows the results obtained

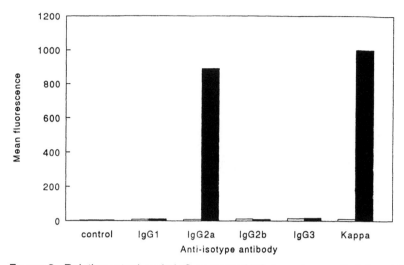

FIGURE 2 Relative cytoplasmic IgG content of NS-1 myeloma cells (□) and MN12 hybridoma cells (■). Incubations were performed with anti-IgG1, IgG2a, IgG2b, IgG3, and anti-κ light chain-specific antibodies. N.C. represents the negative control (only methanol fixation).

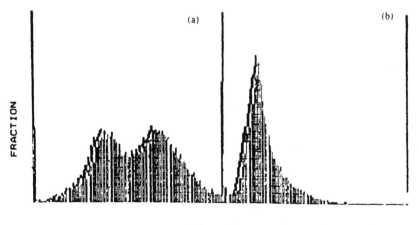

FIGURE 3 Intracellular IgG distribution of the HB-124 hybridoma cells, (a) with Fab fragments of goat antimouse IgG (H + L chain-specific) and (b) negative control of mouse myeloma X-63 cells, (From Ref. 28).

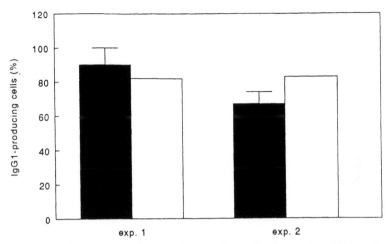

FIGURE 4 Comparison of antibody-secreting cells (■, spot-ELISA) and antibody synthesizing cells (□, FC) in a batch culture of IgG1-producing hybridoma cells.

with FITC-conjugated anti-IgG2a antibodies, used to determine the relative cytoplasmic IgG content. In the same experiments, the antibody activity of the cytoplasmic IgG was determined using the peptide labeled with FITC. The cytoplasmic IgG content increased during the exponential growth phase and decreased rapidly during the decline phase. The two methods agreed to some extent (see Fig. 5A), although the ratio between the relative cytoplasmic IgG content and the antibody activity showed a discrepancy at day 3. This suggests that, at day 3, more heavy chains are synthesized than intact IgG molecules, which bind specifically to the peptide conjugate. Furthermore, an excess of heavy chains could be toxic for the cells [27] and, thereby, affect cell viability, as suggested by the drop in viability at day 4 (see Fig. 5). This can be evaluated using anti-light chain antibodies, as stated previously. All the cells contained cytoplasmic IgG, and a normal gaussian distribution was found, rather than a bimodal distribution, as reported by Altshuler et al. [28]. The mean fluorescence intensity with the anti-IgG2a antibody was higher than that observed with the peptide conjugate (data not shown). This difference is probably caused by the differences of binding properties between the anti-IgG2a–FITC conjugate and the peptide–FITC conjugate; the polyclonal anti-IgG2a–FITC conjugate may recognize several epitopes on one heavy chain of the IgG molecule, whereas the peptide–FITC conjugate binds only to the antigen-binding sites at the Fab parts of the IgG molecule.

The relative membrane IgG content and relative specific membrane IgG content were also determined with the same anti-IgG2a–FITC and peptide–FITC

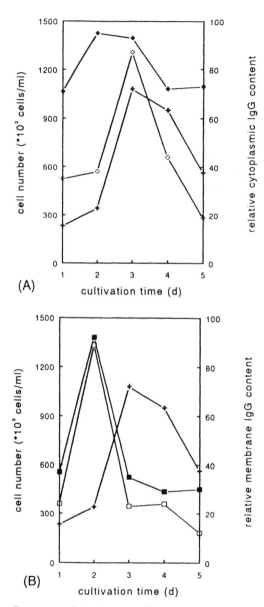

FIGURE 5 Growth curve [(+), viable cells] and relative cytoplasmic IgG content (fluorescence units) of hybridoma cell line MN12 as detected with (A) anti-IgG antibodies (◇) and with the peptide-FITC conjugate (♦) and (B) relative membrane IgG content (fluorescence units) of hybridoma cell line MN12 as detected with anti-IgG antibodies (□) and with peptide-FITC-conjugate (■).

conjugates (see Fig. 5B). The relative membrane IgG content increased during the first day of cultivation, then rapidly decreased until day 3, thereafter remaining constant until the end of cultivation. A decrease in mean surface fluorescence was also observed by Sen et al. [29] and Meilhoc et al. [19]. The relation between mean membrane IgG and specific rate of secretion revealed no correlation ($r = 0.03$; data not shown). There was also no correlation between cytoplasmic IgG content and specific secretion rate ($r = 0.54$; data not shown). In the literature, contradictory results have been reported. Meilhoc et al. [19] also demonstrated a lack of correlation ($r = 0.003$) between specific secretion rate and mean surface IgG content. In contrast, Sen et al. [29] showed a better, although poor, correlation ($r = 0.74$). These observations suggest that any correlation between secretion and surface IgG content may vary between hybridoma cell lines.

The decrease in membrane-associated IgG during cultivation could be caused at the cellular level. The syntheses of surface IgG and secreted IgG may take place by different pathways. Heavy chains that are secreted and those that are incorporated in the cell membrane are encoded by different mRNAs [30, 31]. Thus, a decease in the rate of synthesis of surface IgG does not necessarily affect the rate at which IgG is secreted. Another explanation for the decrease of the surface IgG content may be the physical shear stress caused by agitation of the cell suspension. Surface IgG, which is anchored to the cell membrane by a hydrophobic anchor sequence [31], may be stripped from the cell membrane by these shear forces.

C. Continuous Culture

The growth, cytoplasmic and membrane IgG contents of the MN12 cell line in a continuous culture system are shown in Figure 6. The relative cytoplasmic antibody content showed a more fluctuating course during the first 700 h of culture. At 1200 h of culture, the relative cytoplasmic antibody content showed a more constant pattern. For the first 700 h the cytoplasmic antibody-positive cells (> 95%) consisted of one population. After 1200 h, two populations were observed, one with a relatively high and one with a relatively low cytoplasmic antibody content. Heath et al. [32] also found distinct cytoplasmic antibody content histograms, not only as a function of serum concentration, but also as a function of the culture age. They also described a higher antibody secretion rate in decreasing serum concentrations. Ozturk and Palsson [33] reported the appearance of two populations, one with a high and one with a low relative cytoplasmic antibody content, as they adapted the hybridoma cells to a lower serum concentration. With the appearance of the two populations, a decrease in antibody secretion rate was also observed. This was in contrast with our data and

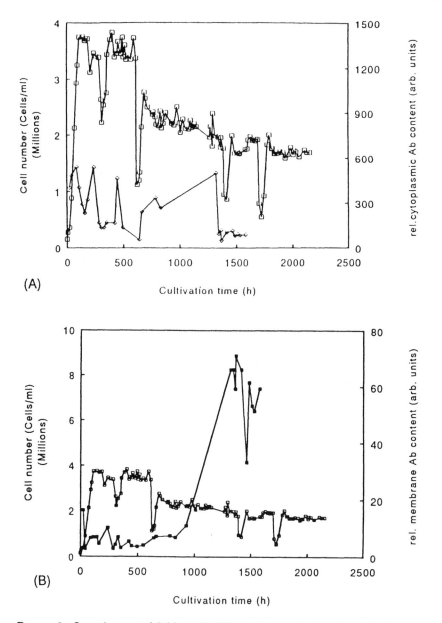

FIGURE 6 Growth curve (viable cells (□)), relative cytoplasmic antibody content (◊) of (B) hybridoma cell line MN12 and (A) relative membrane antibody content (■), in a homogeneous continuous culture system.

those of Heath et al. [32], since, for us, the antibody secretion rate remained constant.

The relative membrane antibody content was low and showed a more constant pattern than did the relative cytoplasmic antibody content during the first 700 h of culture. After 1200 h of culture two populations of membrane antibody-containing cells were observed, each of which contained more antibody on their surface than during the first 700 h of culture.

D. Continuous Perfusion Culture

The growth, the proportion of antibody-secreting cells, and the cytoplasmic IgG content of the MN12 cell line in a continuous perfusion culture system are shown in Figure 7. The proportion of antibody-producing cells, as determined with an isotype-specific spot-ELISA, did not change during the culture period. This indicates that the occurrence of a second steady state did not affect antibody production. The antibody production per cell, expressed both on the basis of the viable cells and on the basis of the percentage of antibody-secreting cells, remained constant throughout the cultivation at a level of 6 µg/10^6 cells per day. The percentage of antibody-producing cells, as determined by FC analysis, was 95% throughout the culture. These results are in good agreement with the number of antibody-secreting cells as determined with the spot-ELISA. This indicates that all antibody-producing cells secrete their IgG into the culture fluid. The relative cytoplasmic IgG content fluctuated during the first exponential growth, the first steady state, and the second exponential growth phases. During the second steady state, the relative cytoplasmic IgG content showed a more constant course. The fluctuating course suggests that, initially, intracellular IgG is accumulated, followed by its secretion into the culture fluid. It is generally assumed that secretory proteins produced by eukaryotic cells are secreted either directly (constitutive secretion) or by secretory granules (regulated secretion) after stimulation [34]. It is also possible that both secretory pathways exist in the same cell [35]. This could mean that, during the first exponential growth phase, the first steady state, and the second exponential growth phase, MN12 hybridoma cells secrete their IgG, to a large extent, according to the regulated pathway. During the second steady state, the secretion pattern suggests a more constitutive pathway. This change in secretory pathway by the cells may be caused by the change of cellular metabolism from aerobic to a more anaerobic condition.

The growth curve, and the antibody properties of secreting cells and of antibody-synthesizing cells for the RIV6 hybridoma cell line are shown in Figure 8. To enumerate the antibody-secreting cells, an isotype-specific spot-ELISA was performed. The proportion of antibody-secreting cells decreased from 45% at the start of cultivation, to 20% at the end. The relative cytoplasmic IgG content was determined with FC. Initially, there were two populations: one

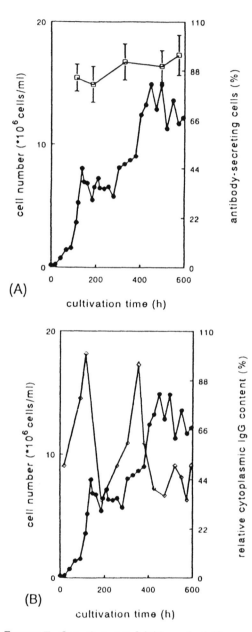

(A)

cultivation time (h)

(B)

cultivation time (h)

FIGURE 7 Growth curve [viable cells (●)], proportion of antibody-secreting cells (□) and cytoplasmic IgG content (◇) of hybridoma cell line MN12 in a homogeneous continuous perfusion culture system.

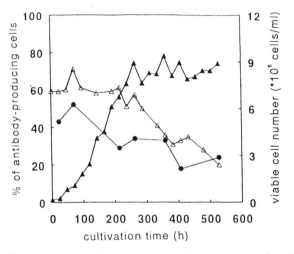

FIGURE 8 Growth curve (▲) and percentage of antibody-producing cells as determined with an isotype-specific spot-ELISA (●) and FC analysis (△).

with a low and one with a high cytoplasmic IgG content (see Fig. 8.). During cultivation the population with the high cytoplasmic IgG content resembling the development of a population with a lower cytoplasmic IgG content.

Heath et al. [32] also reported the existence of two populations of cells: one containing a high and the other a low cytoplasmic IgG content. They showed that this bimodal distribution is dependent on not only the serum content or medium composition, but also on cell aging. However, their experiments were performed in a batch system (spinner flasks) over a period of 5 days. Our experiments were performed in a homogeneous, continuous perfusion culture system and lasted 22 days. Ozturk and Pallson [33] also observed the appearance of a lower antibody-containing population in cells adapted to 1.25% fetal bovine serum (FBS) and to serum-free media. Cells adapted to 20, 10, and 5% FBS consisted of only one, essentially identical, cell population, with a higher intracellular IgG content. This was, in contrast with our observation, during which the cultivation of cells initially in 10% FBS, then reduced to 5% FBS after 240 h of cultivation, resulted in the appearance of a population with a low cytoplasmic IgG content.

Comparison of the results of the FC analysis and spot-ELISA suggests that only the cells with a high cytoplasmic IgG content secrete antibodies, as shown by the FC positive cells in Figure 8. However, there is also a difference between the number of antibody-secreting cells (spot-ELISA) and the number of cells with a high cytoplasmic IgG content (FC analysis). This finding could indicate the presence of a fraction of cells that synthesize cytoplasmic IgG, but are not able to secrete it into the culture fluid. These cells might lack the signals from an intra- or extracellular stimulus to secrete the intracellular IgG [35].

IV. FINAL CONSIDERATIONS

Use of FC analysis facilitates the determination of the stability of monoclonal antibody formation in batch as well as continuous culture systems. By using the appropriate anti-IgG antibodies, FC enables us to measure, in a rather rapid and reliable manner, the cytoplasmic and membrane IgG contents throughout the culture. In combination with other FC measurements of the intrinsic cellular properties of the hybridoma cells, including cell size, viability, DNA (cell cycle), and RNA content, FC seems to be a powerful tool for use in the process control of several pharmaceutical proteins developed in large-scale culture systems. On the other hand, FC can also be applied to the understanding of the fundamental processes involved in the stability or instability of antibody formation (e.g., the appearance of subpopulations of cells with lower cytoplasmic or membrane IgG content). A possible explanation for this phenomenon might be the genetic instability of these hybridoma cells. These cells might have lost the genes encoding the H- and L-chain or may even have lost the complete chromosomes 6 and 12. Recent developments in fluorescence in situ hybridization (FISH) of interphase cells, in combination with whole chromosome-specific probes and FC, would enable the determination of this genetic instability [37]. Moreover this approach could also be used in process control by measuring the of antibody formation during long-term cultures.

REFERENCES

1. Ad Hoc Working Party on Biotechnology/Pharmacy, Notes to applicants for marketing authorizations on the production and quality control of monoclonal antibodies of murine origin intended for use in man, *J. Biol. Stand. 17*:213 (1989).
2. R. Festin, B. Bjorklund, and T. Totterman, Single laser flow cytometry detection of lymphocytes binding three antibodies labelled with fluorescein, phycoerythrin and novel tandem fluorochrome conjugate, *J. Immunol. Methods 126:69* (1990)
3. M. Y. Hoven, L. De Leij, J. F. K. Keij, and T. H. The, Detection and isolation of antigen-specific B cells by the fluorescence activated cell sorter (FACS), *J. Immunol. Methods 117*:275 (1989)
4. J. W. Jacobberger, D. Fogleman, and J. Lehman, Analysis of intracellular antigens by flow cytometry, *Cytometry 7:356* (1986).
5. E. Wenisch, A. Jungbauer, C. Tauer, G. G. Reiter, F. Steindl, and H. Katinger, Isolation of human monoclonal antibody isoproteins by preparative isoelectric focusing in immobilized pH gradients, *J. Biochem. Biophys. Methods 18*:309 (1989).
6. G. R. Hamilton, C. B. Reimer, and L. S. Rodkey, Quality control of murine monoclonal antibodies using isoelectric focusing affinity immunoblot analysis, *Hybridoma 6*:205 (1987).

7. P. Bird, J. E. Cavert, J. Lowe, M. Duggan-Keen, N. G. Forouhi, I. Seppälä, and R. Ling, ELISA measurement of IgG subclass production in culture supernatants using monoclonal antibodies, *J. Immunol. Methods 104*:149 (1987).

8. J. O. Fleming and L. B. Pen, Measurement of the concentration of murine IgG monoclonal antibody in hybridoma supernatants and ascites in absolute units by sensitive and reliable enzyme-linked immunosorbent assays (ELISA), *J. Immunol. Methods 110*:11 (1988).

9. J. M. Coco-Martin, M. Pâques, C. A. M. van der Velden-de Groot, and E. C. Beuvery, Characterization of antibody labelled colloidal gold particles and the applicability in a sol particle immuno assay (SPIA), *J. Immunoassay 11*:31 (1990).

10. J.-P. Aubry, I. Durand, P. De Paoli, and J. Banchereau, 7-Amino-4-methylcoumarin-3-acitic acid-conjugated streptavidin permits simultaneous flow cytometry analysis of either three cell surface antigens or one cell surface antigen as a function of RNA and DNA content, *J. Immunol. Methods 128*:39 (1990).

11. L. W. M. M. Terstappen, H. Meiners, and M. R. Loken, A rapid sample preparation technique for flow cytometric analysis of immunofluorescence allowing absolute enumeration of cell subpopulations, *J. Immunol. Methods 123*:103 (1989).

12. R. Festin, B. Bjorklund, and T. Totterman, Detection of triple antibody-binding lymphocytes in standard single laser flow cytometry using colloidal gold, fluorescein and phycoerythrin as labels, *J. Immunol. Methods 101*:23 (1987).

13. M. G. Wing, A. M. P Montgomery, S. Songsivilai, and J. V. Watson, An improved method for the detection of cell surface antigens in samples of low viability using flow cytometry, *J. Immunol. Methods 126*:21 (1990).

14. K. M. Rigg, B. K. Shenton, I. A. Murray, A. L. Givan, R. M. R. Taylor, and T. W. J. Lennard, A flow cytometric technique for simultaneous analysis of human mononuclear cell surface antigens and DNA, *J. Immunol. Methods 123*:177 (1989).

15. P. Kurki, K. Ogata, and E. M. Tan, Monoclonal antibodies to proliferating cell nuclear antigen (PCNA)/cyclin as probes for proliferating cells by immunofluorescence microscopy and flow cytometry, *J. Immunol Methods 109*:49 (1988).

16. C. V. Clevenger, K. D. Bauer, and A. L. Epstein, A method for simultaneous nuclear immunofluorescence and DNA content quantification using monoclonal antibodies and flow cytometry, *Cytometry 6*:208 (1985).

17. J. L. Dangl and L. A. Herzenberg, Selection of hybridomas and hybridoma variants using the fluorescence activated cell sorter, *J. Immunol. Methods 52*:1 (1982).

18. P. Koolwijk, E. Rozemuller, R. K. Stad, W. B. M. De Lau, and B. J. E. G. Bast, Enrichment and selection of hybrid hybridomas by Percoll density gradient centrifugation and fluorescent-activated cell sorting, *Hybridoma 7*:217 (1988).

19. E. Meilhoc, K. D. Wittrup, and J. E. Bailey, Application of flow cytometric measurement of surface IgG in kinetic analysis of monoclonal antibody synthesis and secretion by murine hybridoma cells, *J. Immunol. Methods 121*:167 (1989).

20. M. Klöppinger, G. Fertig, E. Fraune, and H. G. Miltenburger, High cell density perfusion culture of insect cells for production of baculovirus and recombinant protein, *Production of Biologicals from Animal Cells in Culture* (R. E. Spier, J. B. Griffiths, and B. Meignier, eds.), Butterworth-Heinemann, Oxford, 1991, p. 470.

21. S. E. Merrit and B. O. Palsson, Loss of antibody productivity is highly reproducible in multiple hybridoma subclones, *Biotechnol. Bioeng. 42*:247 (1993)

22. J. S. Kromenaker and F. Srienc, Effect of lactic acid on the kinetics of growth and antibody production in a murine hybridoma: Secretion patterns during the cell cycle, *J. Biotechnol. 34*:13 (1994)

23. J. M. Coco-Martin, P. Koolwijk, C. A. M. van der Velden-de Groot, and E. C. Beuvery, An isotype-specific spot-ELISA for the enumeration of antibody-secreting hybridomas and the determination of isotype switch variants, *J. Immunol. Methods 145*:11 (1991).

24. J. M. Coco-Martin, J. W. Oberink, F. Brunnink, C. A. M. van der Velden-de Groot, and E. C. Beuvery, Instability of a hybridoma cell line in a homogeneous continuous perfusion culture system, *Hybridoma 11*:653 (1992).

25. J. W. Drijfhout, W. Bloemhoff, J. T. Poolman, and P. Hoogerhout, Solid-phase synthesis and applications of *N*-(*S*-acetylmercaptoacetyl) peptides, *Anal. Biochem. 187*:349 (1990).

26. D. Levitt and M. King, Methanol fixation permits flow cytometric analysis of immunofluorescent stained intracellular antigens, *J. Immunol. Methods 96*:233 (1987).

27. G. Köhler, Derivation and diversification of monoclonal antibodies, *EMBO J. 4*:1359 (1985).

28. G. L. Altshuler, R. Dilwith, J. Sowek, and G. Belfort, Hybridoma analysis at cellular level, *Biotechnol. Bioeng. Symp. 17*:725 (1986).

29. S. Sen, W.-S. Hu, and F. Srienc, Flow cytometric study of hybridoma cell culture, correlation between cell surface fluorescence and IgG production rate, *Enzyme Microb. Technol. 12*:571 (1990).

30. F. W. Alt, A. L. M. Bothwell, M. Knapp, E. Siden, E. Mather, M. Koshland, and D. Baltimore, Synthesis of secreted and membrane bound immunoglobulin (heavy chains is directed by mRNAs that differ at their 3' ends, *Cell 20*:293 (1980).

31. S. K. Ghosh, P. Patnaik, and B. Bankert, Expression of μ and g1 membrane forms of immunoglobulin segregate in somatic cell hybrids, *Mol. Immunol. 24*:1335 (1987).

32. C. Heath, R. Dilwith, and G. Belfort, Methods for increasing monoclonal antibody production in suspension and entrapped cell cultures: Biochemical and flow cytometric analysis as a function of medium serum content, *J. Biotechnol. 15*:71 (1990).

33. S. S. Ozturk and B. O. Pallson, Loss of antibody productivity during long-term cultivation of a hybridoma cell line in low serum and serum-free media, *Hybridoma 9*:167 (1990).

34. G. Palade, Intracellular aspects of the process of protein synthesis, *Science 189*:347 (1975).

35. T. L. Burgess and R. B. Kelly, Constitutive and regulated secretion of proteins, *Annu. Rev. Cell Biol. 3*:243 (1987).

36. C. A. M. Van der Velden-de Groot, J. M. Coco-Martin, and E. C. Beuvery, New developments in the cultivation of hybridoma cells in homogeneous continuous perfusion systems, *Dev. Biol. Stand. 71*:45 (1990).

37. H. van Dekken, G. J. van Arkensteyn, J. W. Visser, and J. G. Bauman, Flow cytometric quantification of human chromosome specific repetitive DNA sequences by single and bicolor fluorescent in situ hybridization to lymphocyte interphase nuclei, *Cytometry 11*:153 (1990).

6

Flow Cytometric Studies of Osmotically Stressed and Sodium Butyrate-Treated Hybridoma Cells

STEVE OH
Pall Filtration Pte. Ltd., Singapore

FLORENCE K. F. CHUA
National University of Singapore, Singapore

MOHAMED AL-RUBEAI
The University of Birmingham,
Birmingham, England

I. INTRODUCTION

Among various strategies to increase monoclonal antibody (MAb) production, suppressing cell growth so that hybridomas can divert their energies to MAb production is perhaps one of the more simple and novel strategies. We have shown that eRDF medium (enhanced RDF, which consists of a combination of RPMI:DMEM:Ham's F12 in the ratio 2:1:1, commercially available from Kyokuto Pharmaceutical Industries Co., Japan) is a superior medium for MAb production [1], and that by increasing the osmolarity of a medium and adding sodium butyrate, hybridomas can be stimulated to overproduce MAb [2]. This response has been demonstrated in various cell lines [3]. Cells under such stresses also increase their specific metabolic activity, total RNA content, and the transport of amino acids, whereas growth and total DNA content are concomitantly reduced [4].

In this chapter, we illustrate the use of flow cytometry in examining the relationships between total cellular MAb content, cell size, and cell cycle distribution of hybridomas subjected to environmental stresses. Flow cytometry is a particularly effective tool for population analysis, used here to determine the effects of osmotic pressure and sodium butyrate, as it is able to discriminate the properties of each cell and give an accurate picture of the population profiles. Flow cytometric techniques have been applied to the study of the proliferating capacity of cultured hybridoma cells, changes in cell-specific productivity, and the dependency of cellular MAb content, cell size, and DNA on growth rate under normal conditions [5–10]. Flow cytometry has also been used to determine the MAb levels in cells subjected to transient, but fatal, osmotic stress [10]. Here we have investigated hybridomas that were not adapted as well as those that were adapted to hyperosmotic stress to reveal their cellular responses under such unnatural conditions.

II. METHODS

A. Cell Culture Conditions, Sample Preparation, and Staining Procedures

Cells in normal osmolarity (300 mOsm) were routinely maintained in 10-ml Nunc tissue culture flasks. Hybridoma cells were adapted to higher osmolarity as follows: Cells from normal osmolarity culture were inoculated at a density of 2×10^5/ml into 50-ml tissue culture flasks containing media at 300, 350, and 400 mOsm. Cells were passaged daily in fresh media for a week, at which time the cultures achieved viabilities of over 90%.

At specified time points, 1×10^6 cells were collected from a batch culture in eRDF medium and centrifuged at 1000 rpm with a G6 centrifuge (Beckman, California, USA) to remove the spent medium. Cells were then washed twice with cold phosphate-buffered saline (PBS) and finally fixed in 5 ml of cold 70% ethanol and stored at 4°C before staining for flow cytometry analysis. Dual staining with propidium iodide (PI) for DNA and fluorescein isothiocyanate (FITC)-conjugated goat antimouse IgG (heavy and light chain) allowed simultaneous analyses of cellular DNA and IgG distributions. Fixed cells were washed with PBS, resuspended in 0.1 ml PBS and stained with FITC-conjugated goat antimouse IgG (50 µl of 1:10 dilution) for 30 min at room temperature. Cells were then washed twice with 0.5% Tween 20 in PBS, resuspended in 1 ml PBS, and analyzed in the flow cytometer.

B. Flow Cytometry Analysis

Analysis of the distributions of cell size, cellular antibody, and DNA content were carried out on an Epics Elite flow cytometer (Coulter Electronics, USA) using an excitation power of 15 mW at 488 nm. Forward-angle light scatter was collected through a neutral density filter. Ninety-degree scatter was reflected to one photomultiplier by a 488-nm dichroic mirror and collected through a 488-nm band-pass filter. A 488-nm long-pass filter was used to prevent scattered laser light from reaching the fluorescence detectors. Antibody (green) fluorescence was collected at a second photomultiplier through a 550-nm long-pass dichroic mirror–525-nm band-pass filter combination. Propidium iodide (red) fluorescence was reflected to a third photomultiplier by a long-pass 600-nm dichroic mirror and collected through a 635-nm band-pass filter.

Analyses of DNA and cellular IgG contents were based on selective gating by means of dual multiparameter analysis of peak height and integral PI fluorescence of 10,000–20,000 cells. Gates were constructed to eliminate cellular debris and doublets. Mean fluorescence intensity of cellular IgG was obtained from the "FITC log" histogram. MULTICYCLE (Phoenix Flow Systems) computer analysis of DNA distributions was performed to obtain the percentages of cells in G_1, S, and G_2–M compartments of the cell cycle. The assumptions used by the program are that the DNA contents of cells in G_1 and G_2/M phases are present as two gaussian distributions. The MULTICYCLE software program is based on a polynomial S-phase algorithm with an iterative, nonlinear, least-squares fit.

III. FLOW CYTOMETRIC STUDIES OF STRESSED CELLS

A. Unadapted Cells in High Osmolarity

Osmotic stress can increase the specific productivity of hybridomas [2, 11–13]. Also, sudden changes in osmotic pressure (up to 500 mOsm) caused hybridomas to die, but the specific antibody productivity and cellular antibody content remained high in these dying cells [10]. We have demonstrated that hyridoma cells increase the rate of amino acid transport [4] and, here, we wanted to determine whether these cells had increased cellular MAb levels under high osmotic pressure conditions.

Cells from a control culture at 300 mOsm were inoculated at 2×10^5/ml into media at 300, 350, and 400 mOsm without prior adaptation. Culture preparation was followed exactly as previously described [2]. In short, cells with a viability of greater than 90% were spun down at 1000 rpm for 10 min, and then resuspended with fresh media at the three different osmolarities of 300, 350, and 400 mOsm. All experiments were carried out in 250-ml Duran bottles that contained a liquid volume of 150 ml. Culture bottles were incubated in a 37°C water bath at a shaker speed of 150 rpm to ensure adequate suspension. Samples were taken daily for cell count and MAb analysis. Also, 10^6 cells were collected daily, as described in the methods section for later analysis by flow ctyometry.

Figure 1a shows that, within 24 h, cells in 300-mOsm medium begin to enter the exponential growth phase, peaking at 3×10^6/ml, and the culture is effectively nonviable after 7 days. Cells in the 350-mOsm condition have a longer lag phase of 48 h before rising to a peak of 1.2×10^6/ml, but the culture lasts for 10 days. In the 400-mOsm condition, cell numbers remain at the inoculation density and, then, gradually decline after 5 days of culture. Meanwhile, antibody production in the 300-mOsm culture increases linearly to 150 µg/ml. The cells in 400-mOsm medium, which are heavily suppressed by the high osmolarity, produce only a maximum of 25 µg/ml of MAbs (see Fig. 1b).

If we now look at the changes in cell size and cellular MAb levels during these culture periods, there are clearly physiological changes occurring. Figure 2a shows that, 5 h after cells are seeded into stressful conditions, the cell size is relatively unaffected, but by 24–48 h, cells in 350- and 400-mOsm media are, on average, 10–15% larger than the control. Surprisingly, even though cell numbers continue to increase for up to 100 h, the mean cell size in both 300- and 350-mOsm conditions drops and continues to do so into the death phase. Even for the 400-mOsm culture that is dying, mean cell size remains higher than the control. The shifts are particularly obvious for cells in the 400-mOsm condition after 115 h. The size of the stressed cells is larger than that of the control cells throughout the batch cultures. All the populations retained gaussian distribution profiles throughout the culture period, without any indication of a second population of larger cells being developed (Fig. 3). This is in contrast with other studies when lower serum levels led to the appearance of a population of larger cells [14].

In all of these studies, a similar trend in the relation of MAb population profile to the changes in cell volume can be observed (see Fig. 2b); that is, MAb levels peak at 48 h and decline gradually. The MAb levels are always higher in the 350- and 400-mOsm cultures, compared with that at 300 mOsm. Furthermore, in the 350-mOsm culture, the population of cells remaining after 165 h (when the cells are in the decline phase) continues to show an increasing mean cellular MAb level. Again, a broadening of the population profile was seen (Fig. 4), although the gaussian distribution remained. From the movement of the antibody profiles to the right of the graphs, it is quite clear that these cells contain more cellular MAb. Again, there was no indication of a second subpopulation. But is the average MAb content per unit cell volume also increased? By plotting the value of cellular MAb over cell size, one can obtain an approximate mean volumetric MAb content per unit cell volume (see Fig. 2c). It appears that cells

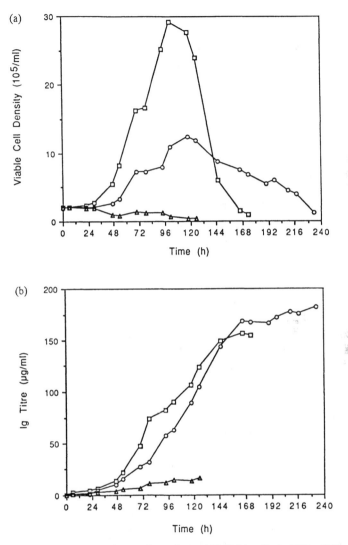

FIGURE 1 Batch culture of unadapted 2HG11 cells in 300-, 350-, and 400-mOsm media: (a) Growth profiles, and (b) immunoglobulin concentration in culture supernatant. Symbols: 300 mOsm (□); 350 mOsm (○); 400 mOsm (△).

in higher-osmotic environments do indeed contain more MAb per unit volume. In 350-mOsm culture, this value tended to rise, even late into the culture. This was the first indication that the cell population remaining after 7 days may have adapted to the environment and become more productive.

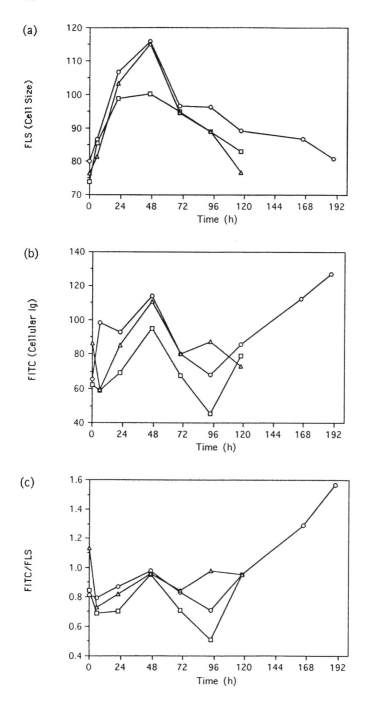

B. Adapted Cells in High Osmolarity

Previous studies of the cellular MAb content of hybridomas focused on monitoring the population of producers when cells were adapted to lower serum [14], and correlation of MAb productivity with cell surface [7] and intracellular MAb [5], all of which were under normal conditions. Here, we are interested in comparing the amounts of cellular MAb under normal and higher osmotic conditions.

Therefore, cells were adapted to high osmolarity by passaging daily in the respective media over 1 week as previously described [2]. Figure 5a shows that rapid growth occurs for all cultures after 24 h. The 300-mOsm culture peaks at approximately 3×10^6/ml, the cells in the 350-mOsm medium peak at marginally lower numbers (2.8×10^6/ml), and those at 400 mOsm reach a maximum of 2×10^6/ml. Figure 5b indicates that production of MAb in the control reaches a maximum of 150 µg/ml. However, the 350- and 400-mOsm cultures produced twice as much, attaining 300 µg/ml, but the MAb concentration in the latter condition reaches this value at a slower rate.

When cellular MAb levels are compared in Figure 5c, once more in all cases the maximum MAb content is reached after 48 h. This may be due to the accumulation of MAb at a faster rate than it is secreted, which would explain why the MAb in the supernatant rises at a slower rate in the early period of the culture. Several researchers have also observed that cellular MAb levels are higher during days 1 and 2 than during days 3 and 4 under normal conditions [5,15]. After 48 h, cellular MAb drops in all cases, but remains highest at 400 mOsm, followed by the 350- and the 300-mOsm cultures. As expected, the adapted cells at higher osmolarities contain more cellular MAb. However, the 400 mOsm culture contains relatively more cellular MAb in the later stages of culture (after 48 h) than does the 350-mOsm medium. Thus, the secretion rate is slower for the former because of the higher amount of MAb retained within the cells. Higher osmolarity selectively inhibits the rate of polypeptide chain initiation and elongation of other cellular proteins more than that of the immunoglobulin H and L chains [16]. The relatively higher amount of intracellular immunoglobulin in cells grown at elevated osmolarity might then be due to the

---

FIGURE 2 Flow cytometric analysis of unadapted cells cultured in 300-, 350-, and 400-mOsm media: (a) batch profiles of forward light scattering (FLS) which measures the cell size; (b) batch profiles of FITC fluorescence, which measures cellular immunoglobulin level; and (c) FITC/FLS, which measures cellular immunoglobulin level per unit cell size. Data represent the mean values of samples at the respective time points. Symbols used: 300 mOsm (□); 350 mOsm (○); 400 mOsm (△).

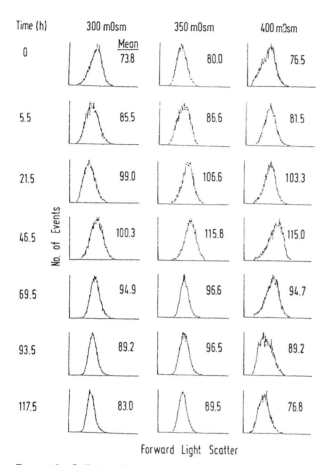

FIGURE 3 Cell size distributions of 2HG11 cells cultured in 300-, 350-, and 400-mOsm media.

higher synthesis of immunoglobulin proteins relative to that of other cytoplasmic proteins.

Although those cells that adapt to low serum media contain lower cellular MAb levels [14], growing cells in enriched media can also enhance cellular MAb levels [15,17]. However, this is the first evidence to show that cells growing in a higher-osmolarity medium produce increased cellular MAb levels. Increased cell surface MAb and intracellular MAb have both been correlated with higher MAb productivity [7,8]. Our data confirms that higher productivity under osmotically stressed conditions can indeed increase the cellular MAb

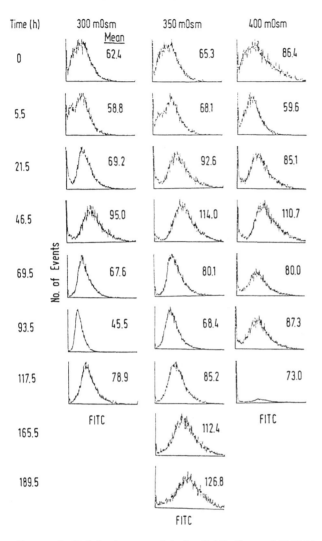

FIGURE 4 Cellular immunoglobulin distributions of 2HG11 cells cultured in 300-, 350-, and 400-mOsm media.

levels. A blocking of the cell cycle with thymidine to reduce growth and increase specific productivity has also been observed [8]. Similarly, we have demonstrated that suppressing cell growth by osmotic stress can also elevate productivity and maintain it over an extended period (provided that the stress is maintained), whereas thymidine blocking of the cell cycle is fatal to cells [8].

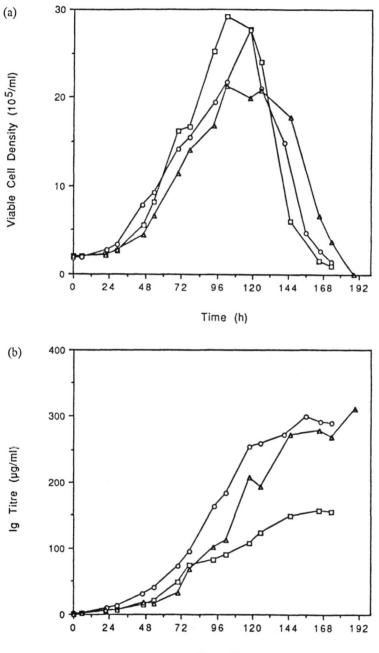

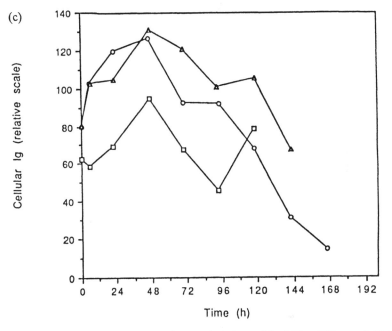

FIGURE 5 Batch culture of 2HG11 cells adapted in 300-, 350-, and 400-mOsm media: (a) growth profiles; (b) immunoglobulin concentration in culture supernatant; and (c) cellular immunoglobulin levels as measured by FITC fluorescence. Symbols: 300 mOsm (□); 350 mOsm (○); 400 mOsm (△).

C. Sodium Butyrate-Treated Cells in High and Normal Osmolarity

In the final series of experiments, we examined the effects of butyrate addition to 300- and 350-mOsm cells on cellular MAb and cell cycle events. It has previously been shown that butyrate can synergistically enhance MAb production in 350-mOsm medium [2]; thus, we were interested in examining the intracellular changes under these conditions. Butyrate was added at 0.1 mM to one set of cultures at 300 mOsm and another at 350 mOsm, while the other set had none. Figure 6a shows that addition of butyrate to 300-mOsm cultures reduces the maximum cell density that can be reached from approximately 3×10^6/ml to 2.4 $\times 10^6$/ml, but both batch cultures last for similar lengths of time. However, when sodium butyrate is added to cells in 350-mOsm culture, there is a longer lag phase (up to 72 h compared with 48 h). The cell densities reach similar maximum densities of 2.2×10^6/ml, but the one with butyrate is prolonged by 2 extra days.

(a)

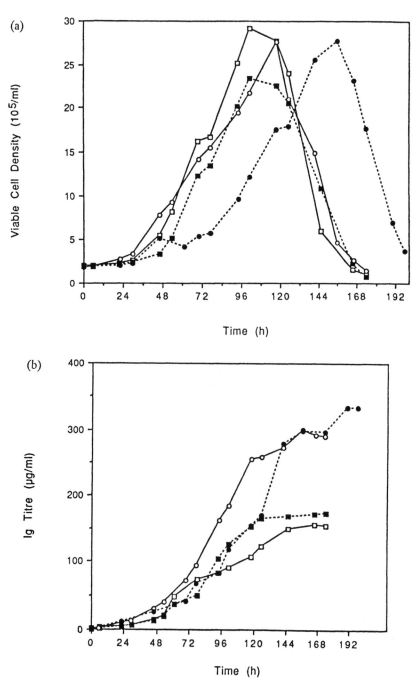

(b)

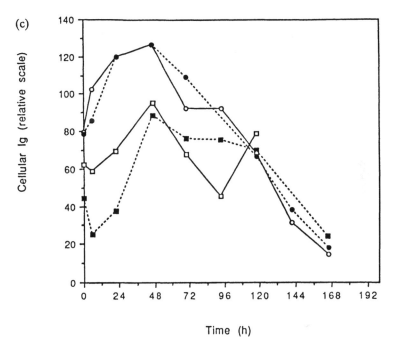

Time (h)

FIGURE 6 Batch culture of 2HG11 cells in 300 mOsm and cells adapted to high osmolarity (350 mOsm), both treated with sodium butyrate: (a) growth profiles; (b) immunoglobulin concentration in culture supernatant; and (c) cellular immunoglobulin levels as measured by FITC fluorescence. Symbols: 300 mOsm (□); 300 mOsm + sodium butyrate (■); 350 mOsm (○); 350 mOsm + sodium butyrate (●).

In terms of MAb levels, Figure 6b indicates that butyrate marginally improves MAb production in the 300-mOsm culture, which reaches 175 μg/ml compared with 150 μg/ml. The rate of production is also greater after approximately 100 h with butyrate. When butyrate is added to the 350-mOsm culture, the MAb production rate is relatively lower than that in the 350-mOsm culture, until after 150 h when MAb concentrations overtake those in the latter, reaching 350 μg/ml.

The total cellular MAb in all the cultures rises to a peak after 48 h as shown in Figure 6c. There is proportionately more total cellular MAb in the 300-mOsm culture without butyrate for the first 48 h, compared with that with the additive, but this is reversed in the latter part of the culture. Thus butyrate addition to the 300-mOsm culture caused more MAb to be retained early in the culture, but less in the later phases of the culture. The cellular MAb is higher in the 350-mOsm cultures with and without butyrate, compared with both the 300-

mOsm conditions. Surprisingly, butyrate addition to the 350-mOsm culture did
not change the cellular MAb levels.

Figure 7 represents data from a repeat experiment that shows a similar
trend of behavior in terms of cell growth. Cell samples were collected at daily
intervals for cell cycle analyses. In this case, 350-mOsm cells were much more
suppressed in growth relative to 300-mOsm cells. Further butyrate addition
slightly suppressed the 300-mOsm cell culture, but when butyrate was added to
350-mOsm culture, it prolonged both the lag phase and the time to achieve peak
cell numbers.

Figures 8a and b show the fractions of G_1, S, and G_2 cells in both the 300-
mOsm culture and the 300-mOsm culture treated with butyrate. The cell cycle
fractions look almost identical. The proportion of G_1 cells started high at 60%,
and then dropped to a minimum of 40% after 24 h before gradually rising to
60% again. The fraction of S phase cells started at a minimum of 20%, and then
rose to a maximum of 50% after 24 h, before dropping to 20% again after 100 h,
this being behavior similar to that observed in normal hybridoma cell culture
[18, 19]. The G_2 fraction is quite insensitive to changes, showing only a slight

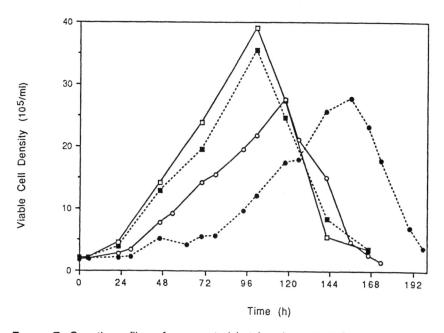

Time (h)

FIGURE 7 Growth profiles of a repeated batch culture of 2HG11 cells in 300
mOsm and cells adapted to high osmolarity (350 mOsm), both treated with
sodium butyrate. Symbols: 300 mOsm (□); 300 mOsm + sodium butyrate (■); 350
mOsm (○); 350 mOsm + sodium butyrate (●).

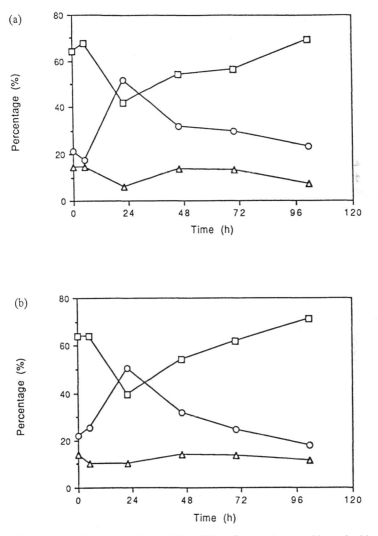

FIGURE 8 Cell cycle analysis of the 300-mOsm cultures with and without sodium butyrate treatment: (a) control 300-mOsm culture, and (b) 300-mOsm + sodium butyrate. Symbols: G_1 phase (□); S phase (○); G_2 phase (△).

dip in value from about 15 to 10% after 24 h before rising again to about 15%. Thus, even though cells are still apparently in the lag phase of growth at 24 h, with no visible increase in cell numbers, about 50% of the population is actually actively synthesizing DNA. During the rest of the culture time, only 20–30% of the cells are doing so.

Figures 9a and b show the fractions of G_1, S, and G_2 cells in the 350-mOsm culture and in the 350-mOsm cultures treated with butyrate. The fraction of the G_1 cells in the 350-mOsm culture (see Fig. 9a) starts at 55% and falls to 40%, before increasing back to 65%. The S phase cell fraction starts at 35%, increasing to 60% after 24 h, and falling to near 20% after 72 h. The proportion

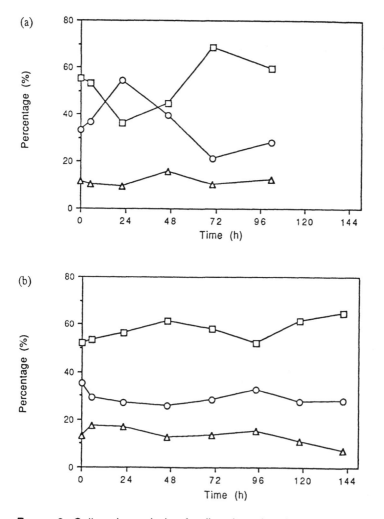

FIGURE 9 Cell cycle analysis of cells adapted to 350-mOsm medium with and without sodium butyrate treatment: (a) 350-mOsm culture, and (b) 350 mOsm + sodium butyrate. Symbols: G_1 phase (□); S phase (○); G_2 phase (△).

of G_2 cells remains at about 10% throughout the culture. This trend is similar to that observed in the lower-osmolarity cultures and to that reported in the previous studies on normal batch cultures. Figure 9b, however, shows that the G_1 cells start at 50% and remain between 50 and 60% throughout the culture period, indicating that a large proportion of the cells in the 350-mOsm culture with butyrate addition remained in the resting phase of the cell cycle. Similarly, the S phase cell fraction stays at only about 30% and G_2 cells remain at 20% throughout the culture. This lack of variation in the cell cycling is indicative of the suppressive nature of butyrate, and this is also reflected in the prolonged lag phase and extended culture period of cells in this condition. Therefore, the additional increase in MAb production levels may be because more cells are in the G_1 cell cycle phase, which has been reported to be the more productive phase in the cell cycle [5].

The suppression of cell growth by addition of thymidine, growth inhibitor [20], and changing the temperature [21] has been attempted, but with limited success. There have been only marginal improvements in final yields, but the cost of the inhibitory chemicals limits their use at larger scales. Butyrate, on the other hand, has been able to suppress cell growth sufficiently to enhance MAb production without causing cell death, and it is not a major cost burden on the process.

IV. CONCLUSIONS

Hybridomas unadapted to high osmolarity are suppressed in growth, although antibody levels are the same as in the control or lower. The changes that occur at the cellular level are increases in cell volume and cellular antibodies, compared with cells in 300-mOsm conditions. When cells are adapted to high osmolarity, suppression of growth is marginal, but both cellular MAb content and the secreted MAb product are elevated by large percentages. The addition of butyrate causes a further increase in secreted MAb levels, but only a marginal increase in cellular levels. Sodium butyrate does not appear to affect cell cycle performance in 300-mOsm cultures, but it suppresses the fraction of S phase cells and increases the fraction of cells in the G_1 phase of the 350-mOsm cultures.

REFERENCES

1. F. Chua, S. K. W. Oh, M. Yap, and W. K. Teo, Enhanced IgG production in eRDF with and without serum, *J. Immunol. Methods 167:*109 (1994).
2. S. K. W. Oh, P. Vig, F. Chua, W. K. Teo, and M. Yap, Substantial overproduction of antibodies by applying osmotic pressure and sodium butyrate, *Biotechnol. Bioeng. 42:*601 (1993).

3. F. K. F. Chua, S. K. W. Oh, and M. Yap, Hyperstimulation of monoclonal antibody production by media selection and high osmolarity stress, *J. Biotechnol. 37*:265 (1994).

4. S. K. W. Oh, F. K. F. Chua, and A. B. H. Choo, Intracellular responses of productive hybridomas subjected to high osmotic pressure, *Biotechnol. Bioeng. 46*:525 (1995).

5. M. Al-Rubeai, A. N. Emery, and S. Chalder, Flow cytometric study of cultered mammalian cells, *J. Biotechnol. 19*:67 (1991).

6. M. Leno, O. W. Merten, F. Vullier, and J. Hache, IgG production in hybridoma batch culture: Kinetics of IgG mRNA, cytoplasmic, secreted- and membrane-bound antibody levels, *J. Biotechnol. 20*:301 (1991).

7. K. L. McKinney, R. Dilwith, and G. Belfort, Manipulation of heterogeneous hybridoma cultures for overproduction of monoclonal antibodies, *Biotechnol. Prog. 7*:445 (1991).

8. M. Al-Rubeai, A. N. Emery, S. Chalder, and D. C. Jan, Specific monoclonal antibody productivity and the cell cycle—comparisons of batch, continuous and perfusion cultures, *Cytotechnology 9*:85 (1992).

9. J. M. Coco-Martin, J. W. Oberink, T. A. M. van der Velden de Groot, and E. C. Beuvery, The potential of flow cytometric analysis for the characterisation of hybridoma cells in suspension cultures, *Cytotechnology 8*:65 (1992).

10. S. Reddy, K. D. Bauer, and W. M. Miller, Determination of antibody content in live versus dead hybridoma cells: Analysis of antibody production in osmotically stressed cultures, *Biotechnol. Bioeng. 40*:947 (1992)

11. K. Oyaas, T. M. Berg, O. Bakke, and D. W. Levine, Hybridoma growth and antibody production under condition of hypersomotic stress, *Advances in Animal Cell Biology and Technology for Animal Bioprocesses* (R. E. Speir, J. B. Griffiths, J. Stephanne, and P. J. Crooy, eds.), Butterworth, London, 1989, p. 212.

12. K. Oyaas, T. E. Ellingsen, N. Dyrset, and D. W. Levine, Utilisation of osmoprotective compounds by hybridoma cells exposed to hyperosmotic stress, *Biotechnol. Bioeng. 43*:77 (1994).

13. S. S. Ozturk and B. O. Palsson, Effect of medium osmolarity on hybridoma growth, metabolism and antibody production, *Biotechnol. Bioeng. 37*:989 (1991).

14. S. S. Ozturk and B. O. Palsson, Loss of antibody productivity during long-term cultivation of a hybridoma cell line in low serum and serum-free media, *Hybridoma 9*:167 (1990)

15. M. Dalili and D. F. Ollis, A flow cytometric analysis of hybridoma growth and monoclonal antibody production, *Biotechnol. Bioeng. 36*:64 (1990).

16. D. L. Nuss and G. Koch, Variation in the relative synthesis of immunoglobulin G and nonimmunoglobulin G proteins in cultured MPC-11 cells with changes in the overall rate of polypeptide chain initiation and elongation, *J. Mol. Biol. 102*:601 (1976).

17. C. Heath, R. Dilwith, and G. Belfort, Methods for increasing monoclonal antibody production in suspension and entrapped cell cultures: Biochemical and flow cytometric analysis as a function of medium serum content, *J. Biotechnol. 15*:71 (1990).

18. O. T. Ramirez and R. Mutharasan, Cell cycle- and growth phase-dependent variations in size distribution, antibody productivity, and oxygen demand in hybridoma cultures, *Biotechnol. Bioeng. 36*:839 (1990).

19. M. Al-Rubeai, S. Chalder, R. Bird, and A. N. Emery, Cell cycle, cell size and mitochondrial activity of hybridoma cells during batch cultivation, *Cytotechnology 7*:179 (1991).

20. E. Suzuki and D. F. Ollis, Enhanced antibody production at slowed growth rates: Experimental demonstration and simple structured model, *Biotechnol. Prog. 6*:231 (1990).

21. G. K. Sureshkumar and R. Mutharasan, The influence of temperature on mouse-mouse hybridoma growth and monoclonal antibody production, *Biotechnol. Bioeng. 37*:292 (1991).

7

Flow Cytometric Analysis of Cells Obtained from Human Bone Marrow Cultures

DAVID BROTT, MANFRED R. KOLLER,
SUE A. RUMMEL, AND BERNHARD PALSSON
Aastrom Biosciences, Inc., Ann Arbor, Michigan

I. INTRODUCTION

In recent years, progress in the elucidation of molecular controls of cell replication and differentiation, on the one hand, and the development of sophisticated cell culture techniques, on the other, has lead to the development of a new field of research called *tissue engineering* [1–4]. The primary goal of this endeavor is to reconstitute functioning human tissue ex vivo, primarily for therapeutic cell transplantation. Many tissues are currently under active research, including bone, cartilage, skin, liver, kidney, and bone marrow [1–4].

Bone marrow has been of particular interest because it is currently being transplanted worldwide, and this practiced therapy would be greatly improved by the ability to expand bone marrow stem and progenitor cells ex vivo [4]. Development of perfusion culture technology for the expansion of human bone marrow cell populations has been the focus of much recent research [5–9]. The resulting cultures have the ability to generate cells of different hematopoietic lineages from mononuclear cells. Although the ability to characterize freshly aspirated bone marrow cells by flow cytometry is highly developed, the corresponding analysis of ex vivo expanded human marrow is not. This chapter will focus on the flow cytometric characterization of human bone marrow that is grown ex vivo.

II. THE BIOLOGY OF BONE MARROW

Bone marrow is a complex tissue comprising many different cell types [4]. Non-hematopoietic support cells, such as fibroblasts, adipocytes, and endothelial cells, make up an important part of the bone marrow microenvironment. These cells form the extracellular matrix and are responsible for directly facilitating hematopoiesis. The hematopoietic process of proliferation and differentiation of cells into the different mature cell lineages is shown in Figure 1. Hematopoiesis is initiated by one group of cells, called pluripotent stem cells, that are capable of proliferating or maturing into cells of any of the hematopoietic lineages. The lineage path that a maturing cell traverses depends on the signals that the cell receives. Various interleukins (IL) and cytokines such as IL-3, IL-6, granulocyte–macrophage-colony stimulating factor (GM–CSF), along with other factors, cause the pluripotent cells to mature through the progenitor stage and into the myeloid lineages. Alternatively, IL-2, IL-3, with IL-4, influence the cells to mature down the lymphocyte lineage.

The study of environmental stimuli on bone marrow culturing is an active area of research. As a cell matures, it loses its capability to proliferate and, thereafter, can mature only through a specific lineage or undergo senescence. For the hematopoietic process to continue, pluripotent stem cells must always be present and continually renew themselves. Therefore, any culture system that will facilitate bone marrow transplantation must have some degree of stem cell self-renewal as well as stem cell proliferation and maturation to expand the culture to yield enough cells for the transplant [9]. A more detailed description of the hematopoietic process can be found in various reports [4,10].

III. FLOW CYTOMETRIC ANALYSIS OF BONE MARROW CELLS

One of the first applications of flow cytometry was the analysis of peripheral blood cells (PBC) [11–14]. The development of techniques for the analysis of the mature cells found in peripheral blood led to research on the use of flow cytometry for the analysis of bone marrow. However, the greater number of cell types displaying different levels of maturity present in bone marrow presented a challenging task for flow cytometrists. Increasing numbers of antibodies for both cell type-specific and stage-specific surface proteins were developed, and clever use of these antibodies in various combinations led to the successful analysis of bone marrow [15–21]. However, since complete analysis requires many differ-

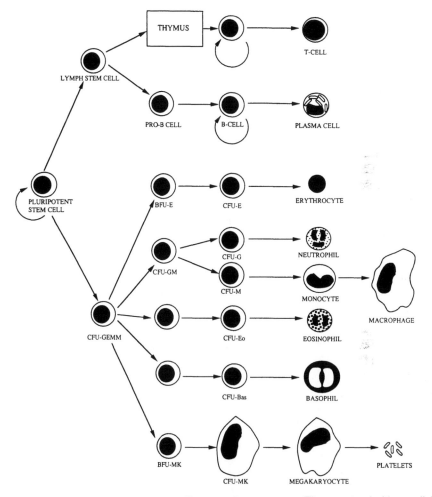

FIGURE 1 The bone marrow cell maturation process. The most primitive cell is the pluripotent stem cell; which can replicate itself or proliferate and mature into one of the multipotent cells. These multipotent cells can proliferate or mature into one of a few different lineage-specific pathways. Further into the lineage, the cells are committed to mature in a single-lineage pathway and can no longer proliferate. (Modified from Ref. 4.)

ent antibody combinations, the simultaneous analysis of different cell populations is severely limited by the number of fluorochromes that can be analyzed by a flow cytometer.

In many instances, hematopoietic tissue culture research and bone marrow transplantation require the quantitation of the most primitive cells. In such situations, maturing cell populations in different lineages can be labeled using a panel of antibodies all conjugated to the same fluorochrome [22]. This approach allows the removal of the large number of mature cells from analysis using a single fluorochrome, while the remaining fluorescent detectors are used to measure the epitopes found on the primitive cell populations of interest. The antibody cocktail used to label the lineage-specific cells is usually referred to as "Lin" and may consist of several different combinations of antibodies. This approach permits using a few antibodies to evaluate the status of immature bone marrow cells. A panel of antibodies that can be used for routine analysis is listed in Table 1. The Lin–FITC cocktail consists of CD3, CD11b, CD15, CD20, and glycophorin-A (gly-A), all conjugated with fluorescein isothiocyanate (FITC) to detect the cells that are lineage-committed.

A. Preparation of Bone Marrow Cells for Flow Cytometric Analysis

Typically, bone marrow aspirates are taken from the iliac crest and collected into heparinized medium. The aspirated bone marrow is then processed by centrifugation through a discontinuous density gradient to remove the mature enucleated erythrocytes and polymorphonuclear granulocytes, thereby yielding a mononuclear cell (MNC) population. Ficoll or Percoll is typically used for the density separation [23]. The low-density MNC are collected from the interface of the

TABLE 1 Useful Antibody Combinations and the Cell Types They Identify[a]

| Antibodies | | |
Primary	Secondary	Cells labelled
IgG$_1$–FITC, IgG–PE	None	Isotype control
CD71–FITC, gly-A–PE	None	Erythroid cells
CD15–FITC, CD11b–PE	None	Monocytic granulocytic
Lin–FITC, IgG–PE, IgG–biotin	SA˙APC	Isotype controls
Lin–FITC, CD38–PE, CD34–biotin	SA˙APC	Pluripotent cell subsets
CD61–FITC, Lin–PE	None	Megakaryocytes
CD19–FITC, CD5–PE	None	Lymphoid cells
Propidium iodide	None	Non-viable cells
Additional antibodies for CD34 subsets		
Lin–FITC, Thy-1–PE, CD34–biotin	SA˙APC	Pluripotent cell subsets
Lin–FITC, c-kit–PE, CD34–biotin	SA˙APC	Pluripotent cell subsets

[a]There are other antibodies that can be used, but these identify the pluripotent cells and the major lineages. Thy-1 and c-*kit* antibodies can be used in conjunction with the Lin cocktail and CD34 antibodies to evaluate subsets of the CD34 population.

Ficoll and heparinized medium, whereas the mature cells pass through the Ficoll and form a pellet at the bottom of the tube. The MNC are ready for flow cytometric analysis or for the initiation of cell culture.

For flow cytometric analysis, the cells are diluted to about 2×10^6 cells per milliliter with phosphate-buffered saline (PBS), containing 0.1% bovine serum albumin (BSA) (staining medium). One-half milliliter is then stained with the appropriate antibodies listed in Table 1 for 15 min at 4°C, and then washed with the staining medium to remove excess antibody. Primary biotinylated antibodies are then stained with a secondary antibody for 15 min at 4°C before washing with the staining medium. This completes the cell-staining process, and the samples are ready for flow cytometric analysis.

The instruments used for acquiring the data presented in this chapter were the Becton-Dickinson (BD; San Jose, CA) FACScan or BD FACSVantage flow cytometers, with appropriate lasers and emission filters for detection of the fluorochromes attached to the antibodies. The FACScan utilizes the standard configuration for the detection of FITC- and phycoerythrin (PE) conjugated antibodies. The FACSVantage uses the 488-nm wavelength of the 5-W argon laser and the 633-nm wavelength from the helium–neon (HeNe) laser, along with five photomultiplier tubes for the measurement of the various fluorochromes. The HeNe laser beam is staggered after the argon beam for separate excitation of the corresponding fluorochromes. The argon laser excites the FITC and PE fluorochromes, and the fluorescent emissions are measured using the standard filter combinations (530 ± 15 nm and 575 ± 13 nm, respectively). Allophycocyanin (APC) is excited with the HeNe laser and the emission is measured using a 660 ± 10-nm and band-pass filter in front of the detector. Since the argon and HeNe lasers are staggered and are not colinear, there is no fluorescence compensation necessary for the APC fluorochrome. Complete lineage analysis including granulocytic–monocytic, erythroid, lymphoid, megakaryocytic–platelet, primitive cells, and cell viability can be obtained with eight stained tubes of cells (see Table 1). Table 1 also includes a few antibodies other than CD38 that can evaluate subsets of the Lin⁻CD34⁺ population.

B. Lineage Analysis of Fresh Bone Marrow

1. Light-Scattering Characteristics

Forward light scatter (FSC) is related to the size of the cell, and side light scatter (SSC) is a measure of cell granularity. As a result, light scatter characteristics can be used to identify different cell types. Lymphocytes are small and agranular and, therefore, have low FSC and SSC (Fig. 2). Monocytes are larger cells, and have more intracellular complexity than lymphocytes and, therefore, have an intermediate level of FSC and SSC. Granulocytes typically have very high SSC owing to granularity, but can vary significantly in size based on the type of

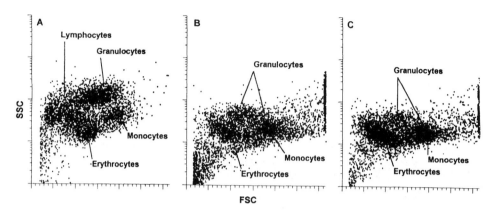

FIGURE 2 Light scatter distributions of bone marrow mononuclear cells using log side light scatter (SSC) and forward light scalter (FSC). (A) Data obtained from freshly aspirated bone marrow following Ficoll processing; (B) data obtained from adherent cells after 14-day culture; (C) data obtained from the nonadherent cells after culturing. Each plot has corresponding populations labeled. The cultured cells (B and C) do not have the lymphocytes labeled because there are so few present after culturing.

granulocyte and the level of activation. Nucleated erythrocytes, or rubricytes, vary in FSC and SSC, based on their maturation stage, and may lie in the same areas on the FSC versus SSC plot as the lymphocytes and monocytes. This population usually has few cells in fresh bone marrow that has been Ficol-processed. The last catagory of cells, the blasts, are few in fresh bone marrow and, therefore, also do not interfere with the myeloid or lymphoid cells. The FSC of the blasts vary, but they are low in SSC. Therefore, light scatter can be used to distinguish cell types into multiple general categories, but more-refined identification requires using other methods, such as fluorescent-conjugated antibodies.

2. Neutrophils and Monocytes

Conjugated antibodies to two well-known surface markers, CD15–FITC and CD11b–PE (Becton Dickinson, San Jose, CA), are used to identify the granulocytic and monocytic lineages (Fig. 3). Further division of these cells into various stages of maturation is described by Terstappen [16]. It is possible to quantify the promyelocytes, myelocytes, metamyelocytes, band, and segmented neutrophils, based on the different level of CD15 and CD11b labeling. Accordingly, these two antibodies not only identify the granulocytic–monocytic cells from the other hematopoeitic cells, but they also allow subclassifying of these lineages into their various maturation stages.

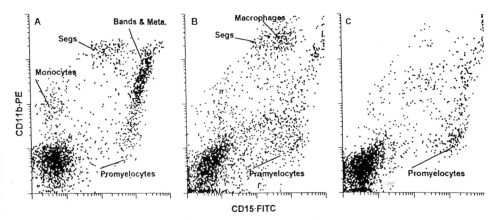

CD15·FITC

FIGURE 3 Bivariate dot plots of bone marrow mononuclear cells stained with CD15–FITC and CD11b–PE to identify the granulocytic and monocytic cells. These populations can be subclassified into their morphological maturation stages [16]. (A) Freshly aspirated bone marrow cells; (B) adherent bone marrow cells after culturing; and (C) nonadherent bone marrow cells after culturing. The identification of the populations are very similar between the fresh and cultured cells, with the difference that macrophages are present in the cultured cells. The macrophages are CD15^{dim} and CD11b bright, which is similar to the segmented neutrophils (Segs). Bands and Meta. (Metamyelocytes) appear as one population.

3. Erythroid Cells

The antibodies CD71–FITC (DAKO, Carpinteria, CA) and gly A–PE (AMAC Inc, Westbrook, ME) identify cells in the erythroid lineage [24] (Fig. 4). High levels of the transferrin receptor (CD71) are present on the very immature erythroid cells, beginning at the burst-forming unit-erythroid (BFU-E) stage (CD71^{2+}gly-A$^-$). As cells progress down this lineage, gly-A becomes expressed, and then CD71 levels decrease as the erythroid cells mature. All cells that are morphologically identifiable as erythroid express gly-A [24].

4. Lymphoid Cells

Any of the antibodies that are used to identify peripheral blood lymphocytes can be used to analyze bone marrow-derived lymphocytes. However, bone marrow aspirates are typically contaminated with peripheral blood, and discriminating between peripheral blood- and bone marrow-derived lymphocytes is currently impossible [25].

The T cells can be detected by using a variety of surface markers, including CD3, CD4, CD5, and CD8, whereas B cells are typically enumerated by

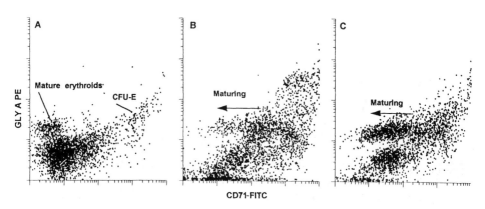

FIGURE 4 Bivariate dot plots of human bone marrow mononuclear cells stained with CD71 and glycophorin-A to identify the erythroid cells. (A) Freshly aspirated bone marrow cells after Ficoll separation; (B) adherent cells after culture; (C) nonadherent cells after culture.

using CD19 or CD20 (Fig. 5). Normal lymphocytes in the bone marrow are rarely positive for both markers. However, in certain pathological situations, such as B-cell chronic lymphocytic leukemia (CLL) or in some cases of non-Hodgkin's lymphoma (NHL), the malignant cells can be positive for both antigens [26, 27] (see Fig. 5). Bone marrow aspirates typically have 5–15% lymphocytes, representing a combination of both B and T cells, and most of the

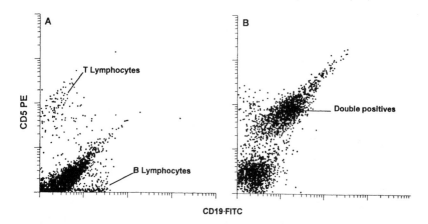

FIGURE 5 Bivariate dot plots of bone marrow mononuclear lymphocytes. (A) "normal" bone marrow cells after Ficoll separation; (B) cells from a non-Hodgkin's lymphoma patient who has double-positive cells.

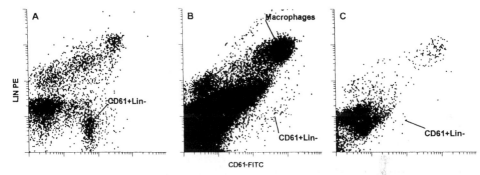

FIGURE 6 Bivariate dot plots of bone marrow mononuclear cells stained for megakaryocytes and platelets. (A) Freshly aspirated bone marrow cells, (B) adherent cells, (C) nonadherent cells, respectively, after 14 days of culture.

lymphoid cells detected are due to peripheral blood contamination of the bone marrow aspirate.

5. Megakaryocytes and Platelets

Cells of the megakaryocytic–platelet lineage express antigens for CD41 and CD61 [19, 20]. To detect cells of this lineage, it is useful to use a cocktail of PE-conjugated antibodies to mature lineage-specific cell markers. Such a cocktail may consist of CD3, CD14, CD19, and gly-A, so that T-, B-, monocytic, and erythroid cells can be removed, as a group, to permit easy detection of megakaryocytic cells (Fig. 6). The FITC-conjugated antibodies to CD61 and CD41 are then used to label the megakaryocytes and platelets. Cell size and granularity (FSC and SSC, respectively) of the CD61+ or CD41+ cells, or both, can be used to distinguish between the platelets and megakaryocytes (Fig. 7). The plot of the fresh aspirate sample shows a large population of platelets and a few megakaryocytes.

Unexpectedly, the megakaryocytes are not off-scale for FSC, as one would expect, owing to their large size. Debili [20] also demonstrated that the small megakaryocytes had cell volumes corresponding to the low to mid-sized range of the entire bone marrow cell population.

DNA analysis is another method used to identify megakaryocytes, which usually are polyploid and have more than 4N of DNA, whereas cells of other lineages that are proliferating will have a maximum of 4N DNA. The DNA analysis requires either fixation of the cells and labeling the DNA with propidium iodide (PI), or visible labeling of the DNA with Hoechst 33342. Both methods have their disadvantages. Fixation prevents culturing of the sorted cells and Hoechst requires UV excitation, which is not available on many instruments. It

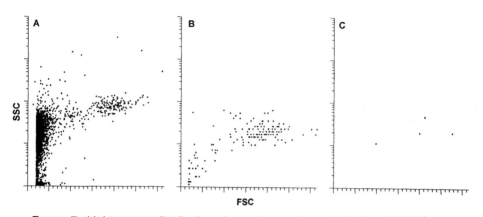

FIGURE 7 Light scatter distribution of megakaryocytes and platelets (CD61+ and Lin-), see Fig. 6. (A) Fresh-aspirate bone marrow mononuclear cells; (B) adherent, and (C) nonadherent bone marrow cells after 14 days of culture.

has been determined that many immature megakaryocytes express CD61, but are not polyploid. In fact, many of the immature megakaryocytes have 2N DNA [19, 20, 28]. Therefore, DNA or DNA with CD61 analysis will not detect early cells in the megakaryocytic lineage.

6. Stem and Progenitor Cells

Perhaps the cells of most interest for bone marrow transplantation and hematopoietic culture analysis are the progenitor and pluripotent stem cell populations [29]. Primitive stem cells can produce progenitor cells for any of the cell lineages, and the resultant lineage-committed progenitor cells have the ability to generate a large number of cells of a particular lineage. The heterogeneous collection of stem and progenitor cells uniformly expresses the CD34 antigen [21]. Because CD34 is not expressed by mature cells of any lineage, but is expressed on stem and progenitor cells of every lineage, it is a stage-specific marker, in contrast with most other antigens, which are lineage-specific. CD34 is expressed by about 0.5–2% of the cells present in the bone marrow MNC population.

To analyze this population, an antibody cocktail against mature cells is used to separate them from the primitive population. Such a cocktail may contain antibodies of CD3, CD11b, CD15, CD20, and gly-A [22]. This FITC-conjugated Lin cocktail of antibodies, and biotin-conjugated CD34 (AMAC, Westbrook, ME) can be used to enumerate the cells that are CD34+ and Lin-. This important cell population contains the hematopoietic stem and progenitor cells (Fig. 8).

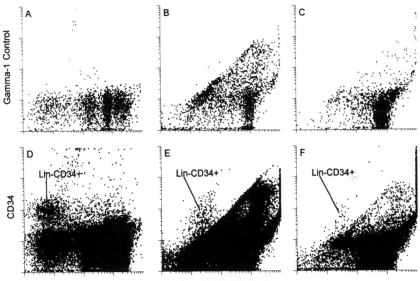

FIGURE 8 Bivariate dot plots of bone marrow cells stained for CD34 and a Lin cocktail for mature cell markers. (A,D) Fresh-aspirated bone marrow mononuclear cells; (B,E) adherent and (C,F) nonadherent bone marrow cells after culture, respectively. (A–C) negative controls; (D–F) CD34 cell population.

The Lin⁻CD34⁺ population can be further subdivided using other antigens, such as CD38 [30], Thy-1 [31], and c-*kit* [32]. The protooncogene c-*kit* is the receptor for stem cell factor (SCF) [33–35], whereas the functions of CD38 and Thy-1 are not yet known. Antibodies to these surface proteins are conjugated with PE (CD38–PE; Becton Dickinson, San Jose, CA; Thy-1, PharMingen, San Diego, CA; and c-*kit*, AMAC Inc.) and therefore, can be used with the Lin–FITC and CD34–biotin–APC stains. These antibodies subdivide the Lin⁻ CD34⁺population differently (Fig. 9). The most immature cells are believed to be c-*kit*⁺, Thy-1ˡᵒ and CD38⁻ [36].

The presence of pluripotent stem cells and lineage-committed progenitor cells in a particular population can be determined by functional assays, such as the colony-forming unit–granulocyte–macrophage (CFU–GM) assay and the long-term culture-initiating cell (LTC-IC) assay [10,37]. The CFU–GM assay detects committed progenitor cells that give rise to a colony of mature cells within a 2-week culture period. These cells are believed to be important for rapid elevation of circulating leukocytes in vivo after bone marrow transplant, but do not result in long-term engraftment [38,39]. The LTC-IC assay detects cells that

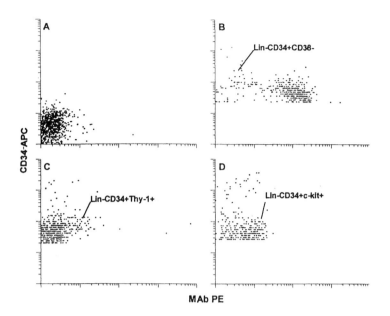

FIGURE 9 Bivariate dot plots of Lin⁻CD34⁺ subsets from fresh-aspirated bone marrow mononuclear cells. (A) isotype controls for both CD34 and its subsets. (B,C,D) CD38, Thy-1, and c-*kit* antibody-labeled subsets of Lin⁻CD34⁺ cells, respectively. The populations of interest are labeled.

can be maintained in growth factor-free culture on preformed stroma for 5–8 weeks and, subsequently, give rise to progenitors (such as CFU–GM) after growth factor stimulation. Therefore, LTC-IC are more primitive cells, and the heterogeneous LTC-IC population is believed to contain cells responsible for long-term in vivo repopulation [37]. The functional characteristics of the subpopulations of the Lin⁻CD34⁺ cells can be determined by sorting the cells into these functional assays.

With the major cell populations identified, it is important to determine the normal ranges for each of these populations. Figure 10 shows the typical values for fresh bone marrow samples. More than 99% of the cells are enucleated red blood cells. After density separation, there are only about 13% erythroid cells and 62% myeloids and only 1.5% Lin⁻CD34⁺ cells. The Lin⁻CD34⁺ cells can be subcategorized by at least three different antibodies; CD38, c-*kit*, and Thy-1. The percentages of subsets have not been verified to correlate with LTC-IC or CFU–GM populations.

A method or antibody combination needs to be developed that correctly predicts the LTC-IC content of a sample. There are several inherent limitations

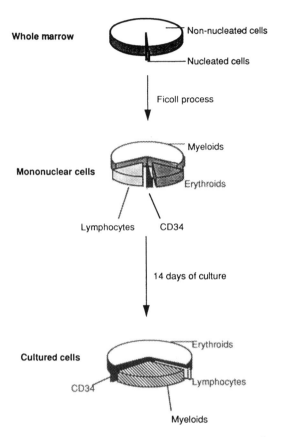

FIGURE 10 The pie chart shows a representative distribution of the cell populations of bone marrow mononuclear cells before and after culturing with IL:3, GM-CSF, SCF, and Epo.

to the quantitation of small populations. A method to overcome these limitations is needed to analyze a sufficient number of cells so that more than 1000 Lin$^-$CD34$^+$ cells will be counted; and lower the detection of false-positive values of the isotype control sample [40]. The goal of subdividing the Lin$^-$CD34$^+$ population is to accurately and consistently measure a value that correctly predicts the LTC-IC content.

IV. CELLS FROM BONE MARROW CULTURES

In recent years, important advances have occurred in the development of ex vivo cultures of human bone marrow. A prolific ex vivo human hematopoietic culture system can be used for various analytical and clinical purposes [4]. Most significantly, a transplantable dose of tumor-free hematopoietic stem and progenitor cells obtained from a small aspirate is desired to improve clinical bone marrow transplantation procedures. The inoculum for such culture systems typically comprises MNC (the flow cytometric analysis of which was described in the last section), although success has recently been reported with the use of whole bone marrow [41] and CD34 selected cells.

After processing and analyzing the bone marrow MNCs, they can be cultured for expansion, and then infused back into the patient. Many research groups are able to generate numerous mature bone marrow cells from purified stem and progenitor cells (the CD34$^+$ population) [42–45], but these cultures result in a decrease of the pluripotent stem cell population [46–48]. The most successful approach yet reported employs a perfusion bioreactor system. This system supports the endogenous formation of stroma from a MNC inoculum, and this ability leads to expansion of the LTC-IC population [8, 49]. It also allows the expansion of the immature and mature cells alike. After culture, which typically lasts 12–14 days, the cells are harvested in three fractions [8]; nonadherent cells come off with gentle agitation, the pseudoadherent cells are harvested using a saline wash, whereas adherent cells require trypsinization. The nonadherent and pseudoadherent cells are typically pooled. The harvesting procedure thus results in two cell populations; adherent cells and nonadherent cells. These cells can be analyzed on the flow cytometer separately or pooled together as one sample. The composition of the cell populations produced can be varied by using different culture conditions.

Cell staining and preparation of cultured cells for flow cytometric analysis is essentially the same as those for freshly aspirated bone marrow cells. However, the various antibodies for analysis must be tested on an individual basis to ensure labeling of the desired cells after culturing and trypsinization.

V. FLOW CYTOMETRIC ANALYSIS OF CULTURED CELLS

The procedures described in the following selection were developed for cultured MNCs, but whole bone marrow should be analyzable on the flow cytometer in the same fashion. The problem with whole bone marrow is that the major population of cells before and after culture are nonnucleated erythrocytes. Just like whole blood lymphocyte analysis, the erythrocytes interfere with the analysis of the nucleated cells. The sample runs slower, because the erythrocytes are counted by the instrument. The erythrocytes of a sample can be lysed. However, one needs to validate that all of the nucleated bone marrow cells remain following the cell lysis procedure or that the distributions of these cells are not affected.

After culture, some cells, including the macrophages, have high autofluorescence and this causes the fluorescence distributions of both the isotype controls and antibody-stained samples to appear different from the fresh bone marrow samples. This autofluorescence is observed in both the FITC and PE fluorescence channels. Figure 11 has the isotype controls of fresh and cultured bone marrow cells. The fresh bone marrow, as expected, appears in the lower-left corner of the histogram. However, the FITC–PE plot of cultured cells has a large tail extending from the lower-left corner toward maximum dual fluorescence. This tail makes it impossible to distinguish between autofluorescent and double-positive cells. There are several approaches to circumvent this problem. For any sample that might have double positives, one of the antigens could be labeled with APC, or the large macrophages could be gated out of the analysis. APC is excited with a 633-nm helium–neon laser and autofluorescence is basically nonexistent when cells are excited at 633, rather than 488 nm. The macrophages become highly autofluorescent during culturing and cause the increased isotype control fluorescence.

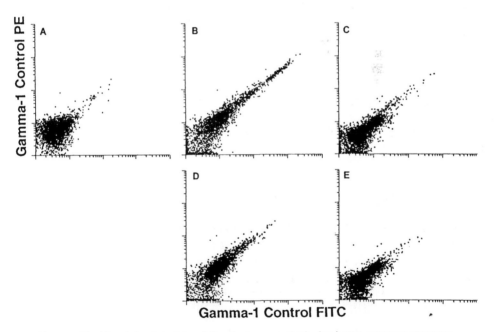

FIGURE 11 Bivariate dot plots of the isotype controls for bone marrow mononuclear cells. (A-C) Based on all cells; (D,E) all cells except the large macrophages. (A) Fresh-aspirated bone marrow; (B,D) adherent cells, and (C,E) nonadherent cells after 14 days of culture.

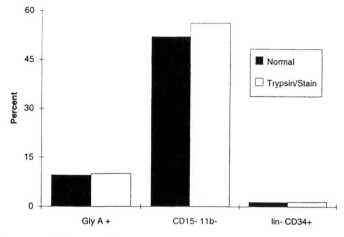

FIGURE 12 The effect of trypsin on antigen expression. Fresh-aspirated bone marrow was stained before and after trypsin treatment to determine if trypsin alters the expression of various antigens.

A. Some Controls

Before analyzing the cultured cells, a few control experiments need to be performed.

1. Effect of Trypsin

To examine the possibility that trypsin alters antigen expression, fresh bone marrow cells were stained before and after trypsin treatment (Fig. 12). Antigen expression was equivalent before and after trypsin treatment. The only time antigen expression was observed to decrease from trypsin treatment was when there was cell clumping. These results demonstrate that the harvesting procedure for adherent cells will most likely not affect the labeling of the cells, and any differences observed between fresh and cultured cells are not due to trypsin treatment.

2. Storage

Often it is not possible to analyze a sample immediately following cell harvest. Consequently, it was necessary to determine the best conditions for sample storage and maximum storage time. Bone marrow aspirates and cultured bone marrow cells are different; therefore, the storage conditions may be different. Fresh bone marrow MNC stored on a Labquake shaker (Labindustries Inc, Berkeley, CA) at 4°C, in medium containing serum, remain viable for at least 72 h (Fig. 13). The labeling of cellular antigens CD71 and gly-A decreased over the 72 h, but the initial level of erythrocytes was low and therefore the decreased

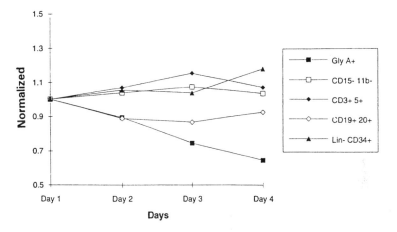

FIGURE 13 Evaluation of storage conditions for fresh-aspirated bone marrow. Fresh bone marrow cells were stored at 4°C on a Labquake shaker (Labin-dustries Inc, Berkeley, CA) and evaluated with antibodies over 4 days.

values were within instrument reproducibility for a sample. Levels of CD15–CD11b and Lin–CD34, remained constant for the 72 h. This enables batching of the samples for optimal use of the instrument and operator's time.

Expanded bone marrow cells require different storage conditions. These cells when stored at 4°C sometimes clump. When clumping occurs, the percentage of viable cells and Lin⁻CD34⁺ cells drop. However, when the cells do not clump, the viability remains high and the percentage Lin-CD34+ remains constant for several days (data not shown). Samples stored at room temperature do not clump, but the storage medium is the major concern. Figure 14 shows that the viability and antibody expression remains constant for 3 days without any medium exchange.

B. Light Scattering Characteristics

Primary human cells typically change in appearance after culture, and this holds true for bone marrow cells. Cultured bone marrow cells (particularly macrophages and granulocytes) appear foamy, with a more irregular shape under the microscope, and have a different size and granularity (FSC and SSC) on the flow cytometer as compared with freshly aspirated cells (see Fig. 2). These changes are partly due to the growth factors used in culture (e.g., GM-SCF), which result in activation of these cells. Therefore, the identification of cultured cells by light scatter is more difficult than the identification of fresh bone marrow MNC (see Fig. 2).

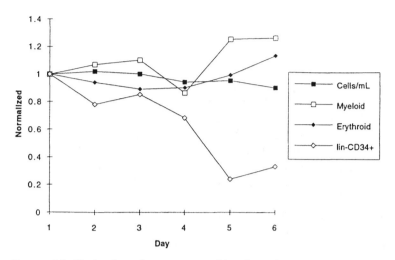

FIGURE 14 Evaluation of storage condition for cultured bone marrow. Cultured bone marrow cells were stored at room temperature and evaluated with antibodies over 6 days.

C. Neutrophils and Monocytes

The expression pattern of CD15 and CD11b on cultured cells is similar to that on freshly aspirated cells. However, the percentages of the different cell types can vary, and the composition of the adherent and nonadherent fractions is usually different. This cell composition can be controlled, to some degree, by manipulation of the culture condition[6]. Figure 3 shows the composition of cells cultured with four growth factors; erythropoietin (EPO), IL-3, GM-CSF, and SCF. This growth factor combination was used in the experiments described in the following, unless otherwise specified.

The adherent cell population is typically rich in segmented neutrophils, metamyelocytes, and bands with some promyelocytes. The nonadherent population is composed of many promyelocytes, and some cells at the myelocyte through band stages, but only a few segmented neurophils. This result indicates that, even though the size and granularity of these cells is altered after culture by flow cytometry, the antigen expression for the granulocytic and monocytic lineages are the same.

D. Erythroid Cells

The expression pattern of CD71 and gly-A on cultured cells is similar to that of MNC (see Fig. 4), except there is a greater prevalence of mature erythroid cells in the cultured samples. The separation process that yields the MNCs removes the mature erythroid cells, and these cells are then regenerated during culture. After ex vivo expansion, the adherent cell population has a significant number of $CD71^+gly-A^+$ cells and $CD71^{dim}gly-A^+$ cells. This population consists of immature erythroid cells, including CFU-Es. In contrast, the nonadherent cell population has a larger subset of mature erythroid cells that are $CD71^-gly-A^+$.

As described by Rogers et al. [24], the maturing erythroid cells follow a specific pathway of CD71–gly-A expression. It was shown, by adding Epo to the cultures, that the earliest committed erythroids, the BFU-E, are $CD71^{2+}$ and gly-A^-. As these cells mature they become gly-A^+, and with further maturation the cells lose CD71 expression. Therefore, this staining combination also allows evaluation of maturation phases of the erythroid cells similar to CD15 and CD11b for the myeloid lineage.

The cultures support maturation of erythroid cells to the mature enucleated red blood cells. Starting with only a few cells in the erythroid series in the inoculum, the cultures have the ability to support erythropoiesis throughout the entire maturation process to yield mature enucleated red blood cells.

E. Lymphoid Cells

With the standard mixture of growth factors (EPO, IL-3, GM-CSF, and SCF) perfused human hematopoietic cell cultures do not generate lymphoid cells. In fact, under these conditions, the number of lymphoid cells steadily decreases over time in culture [41] (Fig. 15). Before culture, the samples have about 5–15% of B and T cells, and after culture, they have 0–1% of T cells, and typically

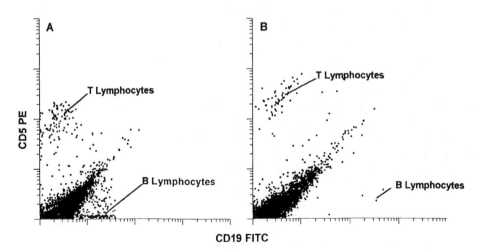

FIGURE 15 Bivariate dot plots of bone marrow mononuclear cells to identify the lymphocytes. (A) Fresh bone marrow; (B) after 14 days of culture.

no B cells. Even when the total cell number after expansion is taken into consideration, the lymphocytes are being depleted from the culture system.

F. Megakaryocytes and Platelets

The appearance of the megakaryocyte and platelet lineage is also similar between cultured cells and the cells inoculated (see Fig. 6). The main difficulty is associated with the identification of the megakaryocytes. After culture, many nonplatelets and nonmegakaryocytes appear CD61$^+$. When dual labeled with the Lin cocktail of antibodies, some cells are positive for both CD61–FITC and Lin-PE. Figure 6 shows a large population of CD61$^+$Lin$^+$ cells. When sorted, the double-positive cells are shown to be macrophages. This result means that CD61 by itself cannot identify megakaryocytes in the cultured samples. The CD61$^+$Lin$^-$ population contains megakaryocytes and platelets for fresh aspirate bone marrow, but Figure 7 shows that neither the adherent nor nonadherent cells have unbound platelets, but there are megakaryocytes present. The absence of platelets may be due to their adherence to Lin$^+$ cells. Platelets are known to stick to cells after their activation. This would explain why CD61 antibody labeling by itself will not identify the megakaryocytes, because the platelets, which are also CD61$^+$, are bound to CD61$^-$ cells.

G. Stem and Progenitor Cells

The percentage of Lin$^-$CD34$^+$ cells decreases during culture; however, their absolute numbers remain constant or increase. Since the percentage of the immature cells is low, it is difficult, and sometimes impossible, to accurately quantitate subsets of these cells using antibodies to CD38, Thy-1, or c-*kit*. However, these antigens are still present on cells after culturing (Fig. 16).

A notable difference between cultured cells and inoculated cells is the percentage of cells that are Lin$^-$CD34$^+$. The nonadherent cultured cells have very few of these cells, but the inoculum and cultured adherent cells can have 0.5–2% Lin$^-$CD34$^+$ cells. The increase in the mature cells demonstrates that these cultures are replicating CD34$^+$ pluripotent cells that mature into the various lineages, whereas the increase in absolute Lin$^-$CD34$^+$ cells indicates an increase of pluripotent cells that remain at this stage of maturation.

H. Analysis of Cell Cycle Status and
Replication History

Since bone marrow cultures result in the expansion of primitive cells [8, 49], they offer a unique model for the study of primitive cell replication. The indelible membrane dye PKH26 (Sigma, St. Louis, MO) can be used to determine the replication history of cells in culture [50]. This dye labels all cells and is equally partitioned between the two daughter cells formed by cell divison. Therefore, the

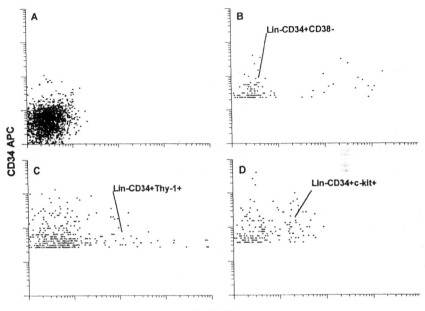

MoAb PE

FIGURE 16 Bivariate dot plots of Lin⁻CD34⁺ subsets of bone marrow mononuclear cells after 14 days in culture. (A) cells labeled with isotype controls; (B–D) Lin⁻CD34⁺ cells stained with CD38, Thy-1, and c-*kit* antibodies, respectively. Cells of interest are labeled.

fluorescence intensity of PKH26 is inversely proportional to the number of divisions that a cell has undergone since the initial staining.

The Lin⁻CD34⁺ cells were labeled with PKH26 and then cultured for several days to determine whether the primitive cells that divide retain the Lin⁻CD34⁺ phenotype, or if cell proliferation causes a cell to mature into Lin⁺ or CD34⁻ cells (Fig. 17). Figure 17C shows hetergeneous PKH26 levels for cultured Lin⁻CD34⁺ cells, indicating the cells are dividing at different rates. When the cells that remain Lin⁻CD34⁺ (see Fig. 17D) are compared with the negative and positive control samples (see Fig. 17A and B, respectively), the replication history of these cells clearly becomes delineated. This subpopulation has clearly undergone multiple rounds of replication and remained at the immature phenotype. Since the culture was inoculated with Lin⁻CD34⁺ cells, PKH26 shows that these cells are capable of proliferating and maturing into the various lineages as well as regenerating the Lin⁻CD34⁺ population. These data are consistent with the measured self-renewal of LTC-IC in perfusion systems [8,49].

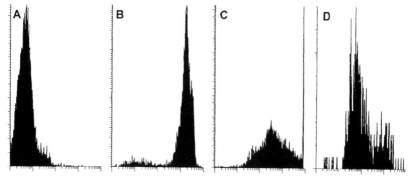

PKH26

FIGURE 17 Histograms of CD34+ bone marrow mononuclear cells labeled with PKH26. Lin⁻CD34+ cells were stained with PKH26 and then cultured for 4 days. (A) Cells not stained with PKH26, but fixed to show the fluorescence intensity of cells if they divided enough to deplete PKH26 levels. (B) Cells stained for PKH26 on day 1 and then fixed. This determines the maximum level of PKH26 for the cell. (C) All cells after 4 days of culture; (D) the PKH26 level of the cells that remained Lin⁻CD34+ after 4 days of culture.

Since PKH26 does not leak from the membrane after fixation, it can also be used in conjunction with various antibodies that distinguish the cell cycle phase and DNA analysis. For example, Ki-67, cdc2-kinase, or cyclin B1 can be used in conjunction with Lin, CD34, PKH26, and DNA staining. These combinations will determine if the Lin⁻CD34+ population has proliferated and remained primitive. This approach will also allow determination of the specific cell cycle phase of the CD34+ cells. This capability is important for many possible research and therapeutic implications. For example, retroviral-mediated gene therapy is thought to require stem cell division, yet these cells must remain in their primitive state to allow bone marrow engraftment.

VI. SUMMARY

Bone marrow culture technology is in its infancy, and we expect that much more will be learned about optimal culture conditions, cell identification, and cellular proliferation and maturation profiles in the future. Flow cytometry offers a rapid, versatile, and important means to analyze the properties of cultured hematopoietic cells. Exciting research lies ahead to develop the ability to analyze the important cellular processes of replication, differentiation, and even self-renewal. Perfused bone marrow cultures offer a unique model to study basic hematopoietic differentiation and the self-renewal of primitive cells.

ACKNOWLEDGMENTS

The authors wish to thank Drs. Gary Van Zant and Craig Jordan for insightful discussions and Mr. Brian Newsom for expert technical assistance.

REFERENCES

1. R. Skalak and C. F. Fox, *Tissue Engineering*, Alan R. Liss, New York, 1988.
2. J. A. Hubbel, B. O. Palsson, and E. T. Papoutsakis, *Biotechnology and Bioengineering: Special Issues on Tisue Engineering and Cell Therapies*, Wiley-Interscience, New York, 1994.
3. R. Langer and J. P. Vacanti, Tissue engineering, *Science 260*:920 (1993).
4. M. R. Koller and B. O. Palsson, Tissue engineering: Reconstitution of human hematopoiesis ex vivo, *Biotechnol. Bioeng. 42*:909 (1993).
5. R. M. Schwartz, B. O. Palsson, and S. G. Emerson, Rapid medium perfusion rate significantly increases the productivity and longevity of human bone marrow cultures, *Proc. Natl. Acad. Sci. USA 88*:6760 (1991).
6. R. M. Schwartz, S. G. Emerson, M. F. Clarke, and B. O. Palsson, In vitro myelopoiesis stimulated by rapid medium exchange and supplementation with hematopoietic growth factors, *Blood 78*:3155 (1991).
7. B. O. Palsson, S. H. Paek, R. M. Schwartz, M. Palsson, G. M. Lee, S. M. Silver, and S. G. Emerson, Expansion of human bone marrow progenitor cells in a high cell density continuous perfusion system, *Biotechnology 11*:368 (1993).
8. M. R. Koller, S. G. Emerson, and B. O. Palsson, Large-scale expansion of human stem and progenitor cells from bone marrow mononuclear cells in continuous perfusion culture, *Blood 82*:378 (1993).
9. S. G. Emerson, B. O. Palsson, and M. F. Clarke, The construction of high effeciency human bone marrow tissue ex vivo, *J. Cell. Biochem. 45*:268 (1991).
10. H. J. Sutherland, A. C. Eaves, and C. J. Eaves, Quantitative assays for human hemopoietic progenitor cells, *Bone Marrow Processing and Purging* (A. P. Gee, ed.), CRC Press, Ann Arbor, 1991, p. 155
11. L. R. Adams and L. A. Kamentsky, Fluorimetric characterization of six classes of human leukocytes, *Acta Cytol. 18*:389 (1974).
12. J. A. Steinkamp, A. Romero, and M. A. Van Dilla, Multiparameter cell sorting: Identification of human leukocytes by acridine orange fluorescence, *Acta Cytol. 17*:113 (1973).
13. H. M. Shapiro, E. R. Schildkraut, R. Cubelo, C. W. Laird, R. B. Turner, and T. Hirschfeld, Combined blood cell counting and classification with fluorochrome stains and flow instrumentation, *J. Histochem. Cytochem. 24*:396 (1976).
14. H. M. Shapiro, Fluorescent dyes for differential counts by flow cytometry: Does histochemistry tell us much more than cell geometry? *J. Histochem. Cytochem. 25*:976 (1977).
15. L. W. M. M. Terstappen, Z. Hollander, H. Meiners, and M. R. Loken, Quantitative comparison of myeloid antigens on five lineages of mature peripheral blood cells, *J. Leukocyte Biol. 48*:138 (1990).

16. L. W. M. M. Terstappen, M. Safford, and M. R. Loken, Flow cytometric analysis of human bone marrow III. Neutrophil maturation, *Leukemia 4*:657 (1990).

17. L. W. M. M. Terstappen, S. Huang, and L. J. Picker, Flow cytometric assessment of human T-cell differentiation in thymus and bone marrow, *Blood 79*:666 (1992).

18. M. R. Loken, V. O. Shah, K. L. Dattilio, and C. I. Civin, Flow cytometric analysis of human bone marrow: I. Normal erythroid development, *Blood 69*:255 (1987).

19. D. Zucker-Franklin, J. S. Yang, and G. Grusky, Characterization of glycoprotein IIb/IIIa-positive cells in human umbilical cord blood: Their potential usefulness as megakaryocyte progenitors, *Blood 79*:347 (1992).

20. N. Debili, C. Issaad, J. M. Massé, J. Guichard, A. Katz, J. Breton-Gorius, and W. Vainchenker, Expression of CD34 and platelet glycoproteins during human megakaryocytic differentiation, *Blood 80*:3022 (1992).

21. C. I. Civin, L. C. Strauss, C. Brovall, M. J. Fackler, J. F. Schwartz, and J. H. Shaper, Antigenic analysis of hematopoiesis. III. A hematopoietic progenitor cell surface antigen defined by a monoclonal antibody raised against KG-Ia cells, *J. Immunol. 133*:157 (1984).

22. J. G. Bender, K. L. Unverzagt, D. E. Walker, W. Lee, D. E. Van Epps, D. H. Smith, C. C. Stewart, and L. B. To, Identification and comparison of CD34-positive cells and their subpopulations from normal peripheral blood and bone marrow using multicolor flow cytometry, *Blood 77*:2591 (1991).

23. H. J. Sutherland, P. M. Lansdorp, D. H. Henkelman, A. C. Eaves, and C. J. Eaves, Functional characterization of individual human hematopoietic stem cells cultured at limiting dilution on supportive marrow stromal layers, *Proc. Natl. Acad. Sci. USA 87*:3584 (1990).

24. C. E. Rogers, M. S. Bradley, B. O. Palsson, and M. R. Koller, Flow cytometric analysis of human bone marrow perfusion cultures: Erythroid development and relationship with burst-forming unit-erythroid (BFU-E) (in press).

25. C. I. Civin, M. L. Banquerigo, L. C. Strauss, and M. R. Loken, Antigenic analysis of hematopoiesis. VI. Flow cytometric characterization of My-10-positive progenitor cells in normal human bone marrow, *Exp. Hematol. 15*:10 (1987).

26. A. S. Freedman, Immunobiology of chronic lymphocytic leukemia, *Hematology 4*:405 (1990).

27. M. Schwonzen, C. Pohl, T. Steinmetz, W. Seckler, B. Vetten, J. Thiele, D. Wickramanayake, and V. Diehl, Immunophenotyping of low-grade B-cell lymphoma in blood and bone marrow: Poor correlation between immunophenotype and cytological/histological classification, *Br. J. Haematol. 83*:232 (1993).

28. C. P. Stahl, D. Zucker-Franklin, B. L. Evatt, and E. F. Winton, Effects of human interleukin-6 on megakaryocyte development and thrombocytopoiesis in primates, *Blood 78*:1467 (1991).

29. R. J. Berenson, W. I. Bensinger, R. S. Hill, R. G. Andrews, J. Garcia-Lopez, D. F. Kalamasz, B. J. Still, G. Spitzer, C. D. Buckner, I. D. Bernstein, and E. D. Thomas, Engraftment after infusion of CD34+ marrow cells in patients with breast cancer or neuroblastoma, *Blood 77*:1717 (1991).

30. L. W. M. M. Terstappen, S. Huang, M. Safford, P. M. Lansdorp, and M. R. Loken, Sequential generations of hematopoietic colonies derived from single nonlineage-committed CD34+CD38− progenitor cells, *Blood 77*:1218 (1991).

31. C. E. Müller-Sieburg, K. Townsend, I. L. Weissman, and D. Rennick, Proliferation and differentiation of highly enriched mouse hematopoietic stem cells and progenitor cells in response to defined growth factors, *J. Exp. Med. 167*:1825 (1988).

32. T. Papayannopoulou, M. Brice, V. C. Broudy, and K. M. Zsebo, Isolation of c-kit receptor-expressing cells from bone marrow, peripheral blood, and fetal liver: Functional properties and composite antigenic profile, *Blood 78*:1403 (1991).

33. D. E. Williams, J. Eisenman, A. Baird, C. Rauch, K. Van Ness, C. J. March, L. S. Park, U. Martin, D. Y. Mochizuki, H. S. Boswell, G. S. Burgess, D. Cosman, and S. D. Lyman, Identification of a ligand for the c-*kit* proto-oncogene, *Cell 63*:167 (1990).

34. K. M. Zsebo, J. Wypych, I. K. McNiece, H. S. Lu, K. A. Smith, S. B. Karkare, R. K. Sachdev, V. N. Yuschenkoff, N. C. Birkett, L. R. Williams, V. N. Satyagal, W. Tung, R. A. Bosselman, E. A. Mendiaz, and K. E. Langley, Identification, purification, and biological characterization of hematopoietic stem cell factor from buffalo rat liver-conditioned medium, *Cell 63*:195 (1990).

35. E. Huang, K. Nocka, D. R. Beier, T. Y. Chu, J. Buck, H. W. Lahm, D. Wellner, P. Leder, and P. Besmer, The hematopoietic growth factor KL is encoded by the *Sl* locus and is the ligand of the c-*kit* receptor, the gene product of the *W* locus, *Cell 63*:225 (1990).

36. Y. Gunji, M. Nakamura, H. Osawa, K. Nagayoshi, H. Nakauchi, Y. Miura, M. Yanagisawa, and T. Suda, Human primitive hematopoietic progenitor cells are more enriched in KITlow cells than in KIThigh cells, *Blood 82*:3283 (1993).

37. C. J. Eaves, H. J. Sutherland, C. Udomsakdi, P. M. Lansdorp, S. J. Szilvassy, C. C. Fraser, R. K. Humphries, M. J. Barnett, G. L. Phillips, and A. C. Eaves, The human hematopoietic stem cell in vitro and in vivo, *Blood Cells 18*:301 (1992).

38. M. J. Robertson, R. J. Soiffer, A. S. Freedman, S. L. Rabinowe, K. C. Anderson, T. J. Ervin, C. Murray, K. Dear, J. D. Griffin, L. M. Nadler, and J. Ritz, Human bone marrow depleted of CD33-positive cells mediates delayed but durable reconstitution of hematopoiesis: Clinical trial of MY9 monoclonal antibody-purged autografts for the treatment of acute myeloid leukemia, *Blood 79*:2229 (1992).

39. R. E. Ploemacher, J. P. van der Sluijs, C. A. van Beurden, M. R. Baert, and P. L. Chan, Use of limiting dilution type long-term marrow cultures in frequency analysis of marrow-repopulating and spleen colony-forming hematopoietic stem cells in the mouse, *Blood 78*:2527 (1991).

40. H. J. Gross, B. Verwer, D. Houck, and D. Recktenwald, Detection of rare cells at a frequency of one per million by flow cytometry, *Cytometry 14*:519 (1993).

41. M. R. Koller, I. Manchel, B. S. Newsom, M. A. Palsson, and B. O. Palsson, Bioreactor expansion of human bone marrow: Comparison of unprocessed, density-separated, and CD34-enriched cells, *J. Hematother. 4*:159(1995).

42. D. N. Haylock, L. B. To, T. L. Dowse, C. A. Juttner, and P. J. Simmons, Ex vivo expansion and maturation of peripheral blood CD34+ cells into the myeloid lineage, *Blood 80:*1405 (1992).

43. W. Brugger, W. Möcklin, S. Heimfeld, R. J. Berenson, R. Mertelsmann, and L. Kanz, Ex vivo expansion of enriched peripheral blood CD34+ progenitor cells by stem cell factor, interleukin-1b (IL-1b), IL-6, IL-3, interferon-γ, and erythropoietin, *Blood 81*:2579 (1993).

44. S. Saeland, C. Caux, C. Favre, J. P. Aubry, P. Mannoni, M. J. Pébusque, O. Gentilhomme, T. Otsuka, T. Yokota, N. Arai, K. Arai, J. Banchereau, and J. E. de Vries, Effects of recombinant human interleukin-3 on CD34-enriched normal hematopoietic progenitors and on myeloblastic leukemia cells, *Blood 72*:1580 1988.

45. J. M. Kerst, I. C. Slaper-Cortenbach, A. E. G. von dem Borne, C. E. van der Schoot, and R. H. van Oers, Combined measurement of growth and differentiation in suspension culture of purified human CD34-positive cells enables a detailed analysis of myelopoiesis, *Exp. Hematol. 20:*1188 (1992).

46. H. J. Sutherland, C. J. Eaves, P. M. Lansdorp, J. D. Thacker, and D. E. Hogge, Differential regulation of primitive human hematopoietic stem cells in long-term cultures maintained on genetically engineered murine stromal cells, *Blood 78*:666 (1991).

47. C. M. Verfaillie, Direct contact between human primitive hematopoietic progenitors and bone marrow stroma is not required for long-term in vitro hematopoiesis, *Blood 79*:2821 (1992).

48. H. J. Sutherland, D. E. Hogge, D. Cook, and C. J. Eaves, Alternative mechanisms with and without steel factor support primitive human hematopoiesis, *Blood 81*:1465 (1993).

49. D. J. Oh, B. O. Palsson, and M. R. Koller, Replating of bioreactor expanded human bone marrow results in extended growth of primitive and mature cells, *Cytotechnology* (in press; 1995).

50. P. M. Lansdorp and W. Dragowska, Maintenance of hematopoiesis in serum-free bone marrow cultures involves sequential recruitment of quiescent progenitors, *Exp. Hematol.* 21:1321 (1993).

8

Measurement of Intracellular pH During the Cultivation of Hybridoma Cells in Batch and Continuous Cultures

JEAN-MARC ENGASSER, ANNIE MARC, AND MARC CHERLET
Institut National Polytechnique de Lorraine,
Vandœuvre-lès-Nancy, France

PIERRE NABET AND PATRICIA FRANCK
Centre Hospitalier Régional Universitaire de Nancy,
Nancy, France

I. INTRODUCTION

The pH of the culture medium represents a major operational parameter in mammalian cell cultures. The rates of cellular growth and metabolism and of protein synthesis are strongly dependent on the medium pH which, thus, must be strictly controlled for efficient production processes of cells, viruses, or proteins.

For cell growth, the external pH is usually maintained between 6.8 and 7.2, depending on the cell line [1]; for the production of proteins, the optimal pH may be different [2]. For instance, culture conditions have been described for the production of antibodies during which hybridoma are initially grown at pH 7.0, whereas the subsequent phase of antibody secretion is carried out at a lower pH of 6.8.

An understanding on a more fundamental level of the effect of medium's pH on cellular activities is largely complicated because the intracellular pH (pH_i), which actually regulates the cell metabolism and division, is generally higher than the external pH (pH_e). Differences between external and internal pH, which are in the range 0.1–0.5 pH unit, seem to depend on the cell line, the medium composition, and the growth conditions [3].

Several mechanisms related to the cellular metabolism and to membrane transport have a direct influence on pH_i. Regulation of pH_i is accomplished through an active extrusion of H^+ by a $Na^+–H^+$ antiport localized in the plasma membrane and driven by the energy of the inward Na^+ electrochemical gradient. This exchange of Na^+ and H^+ by the antiport occurs at a rate that depends on the pH_i [4].

According to recent studies, a strong correlation is likely to exist between pH_i and cell proliferation. Usually, an increase in pH_i can be related to an increase in cell activity (glycolysis, or protein or DNA synthesis). Different works agree in saying that pH_i varies throughout the cell cycle, and that increased proliferation (DNA synthesis, mitosis, increase in cell number) is related with intracellular alkalinization. For instance, in different tumor cell lines [4], pH_i was more alkaline during S, G_2, and M phases, and the quiescence appeared to be triggered by a fall in pH_i of 0.2–0.3 pH unit. In murine B lymphocytes [5], an alkalinization of mitogenic-stimulated cells was also observed before their entry into S phase. Other studies, aimed at understanding the action of inhibitory metabolites, such as ammonia, have also suggested an essential role of intracellular pH. Addition of ammonium ions to the medium has been reported to inhibit cell growth by diffusing into the cells and lowering pH_i [6, 7].

More precise investigations of the influences of intracellular pH on cell activities have been made possible by recent progress in the techniques of pH_i measurement based on fluorescent dyes and flow cytometry. Improvements in flow cytometric technology and synthesis of better-performing dyes and fluorescent molecules now allow a fast analysis of each individual cell, yielding substantial quantitative information on a large cell population and on cell-to-cell heterogeneity [8, 9].

So far, pH_i monitoring has been performed mainly on cells suspended in buffer and submitted to different environmental conditions (external pH, osmotic pressure, addition of lactate or ammonia). The pH_i evolution was followed during short periods, not exceeding 1 h [6, 10]. It is only recently that pH_i has been investigated on cells grown in mass cultures, either batch or continuous, to analyze time variations of pH_i during the consecutive culture phases. Studies have also been begun to investigate the effect on pH_i of extended exposure of cells to unfavorable environments.

In its first part, this chapter reviews the techniques available for the measurement of pH_i using flow cytometry. It then describes the protocols for application of the techniques to the study of cells grown in bioreactors. Finally, it gives some illustrations of the monitoring of pH_i during batch and continuous cultures of hybridoma cells in bioreactors.

II. FLUORESCENT INDICATORS FOR INTRACELLULAR pH MEASUREMENT

Current fluorimetric methods for measuring pH_i rely on the use of membrane impermeant dyes. Uncharged nonfluorescent precursors added to the medium diffuse inside cells. Dyes are then cleaved by intracellular esterases from the precursors and, because of their charged nature, the cleaved products are strongly retained inside the cells.

The wavelength or intensity of emission or excitation of the dye is pH-sensitive. Spectral changes are generally monitored by ratiometric techniques, by measuring the ratio of emission intensities at two wavelengths, since ratio techniques have the advantage of making the pH measurement independent of the cell-to-cell variables, such as the intracellular dye concentration. For flow cytometry, it is preferable to use dyes having a shift in the emission spectrum, since it is technologically less complicated to excite with one wavelength and to split the emitted fluorescence into two separate bands.

Three main pH_i indicators have been available for several years: DCH, BCECF, and SNARF. The choice of fluorochrome depends on the instrumentation available and the particular experimental requirements.

A. 2,3-Dicyanohydroquinone

The pH_i indicator (DCH), which was introduced in 1981 by Valet et al. [8], has a shift in emission wavelength and a decrease in emission intensity with increasing pH. It has a pK of 8.0 and an absorbance maximum in the UV range. Although the DCH technique allows high-sensitivity measurements, it has not been widely used because it requires an UV-excitation source and leaks out of cells rapidly.

B. 2',7'-Bis(2-Carboxyethyl)-5,6-Carboxyfluorescein

The most frequently used fluorescent probe has been 2',7'-bis(2-carboxyethyl)-5,6-carboxyfluorescein (BCECF). It is confined relatively specifically to the cytoplasm and is well retained within the cell, so that stable ratio measurements

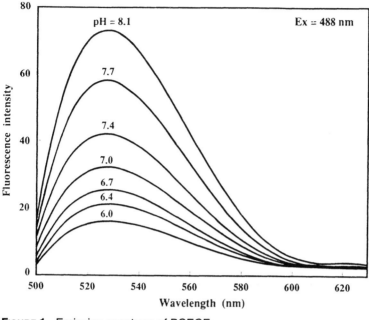

FIGURE 1 Emission spectrum of BCECF.

can be made over at least 2 h. The pK of BCECF is 7.0, which is in the range of expected physiological pH values. The absorption maximum is 500 nm, and the dye can be readily used in flow cytometers equipped with argon ion lasers or with mercury arc lamps. The emission intensity, maximum at 520 nm, increases with increasing pH (Fig. 1). BCECF has been used by measuring the ratio of emission intensities at two wavelengths. A main limitation of BCECF is the relatively reduced shift in emission spectrum as a function of pH. Thus, it offers a lower resolution than DCH for mainly the higher values of pH_i.

C. Carboxy-seminaphthorhodafluor

Carboxy-seminaphthorhodafluor (SNARF-1) belongs to a more recent class of dyes that have a greater pH-dependence shift in emission wavelength (Fig. 2). It is excited with visible argon laser light at 514 nm and presents the advantage of two distinct pH-sensitive emission bands. Its fluorescence ratio (625/575 nm) variation over the pH range is wider than with BCECF, allowing a better sensitivity, especially in the range 7.0–8.0. The leakage from the cells is similar to that of BCECF. It binds to intracellular proteins and this results in a reduced fluorescence intensity [11].

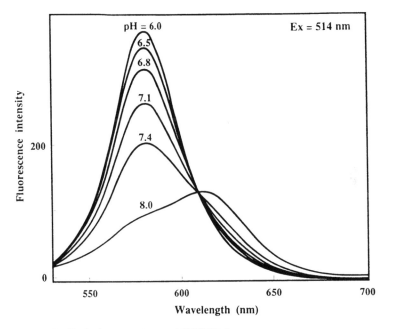

FIGURE 2 Emission spectrum of SNARF-1.

III. PROTOCOL FOR pH$_i$ MEASUREMENTS BY FLOW CYTOMETRY

The protocol described has been developed in our laboratory for the measurement of pH$_i$ in hybridoma cells grown in bioreactors, with BCECF as the fluorescence probe. The cells are sampled at different times during culture and are stained with BCECF before being analyzed by flow cytometry. Calibration curves are established with the same sampled cells. All the steps of sample preparation and calibration must be carefully performed, as the measurement is very dependent on the control of each one step.

A. Cell Loading with the pH$_i$ Indicator

The loading of the cultured cells with the selected dye involves the following steps:

1. Harvest about 10^7 cells, centrifuge them, wash once in 2 ml EBSS buffer (at 20°C and pH adjusted to 7.2), and suspend the cells at a concentration of 10^7/ml in the same buffer. The duration of this initial cell-sampling and washing step at pH 7.2 and 20°C has no significant

effect on the final fluorescence ratio measurement, provided that it does not exceed 120 min. It is normally performed in less than 30 min.

2. To the harvested cells, add 10 µl BCECF, from a stock solution of 1 mM in dimethylsulfoxide, to a final concentration of 10 µM. Staining 10^7 cells with 10 µM BCECF results in a significant and homogeneous fluorescence of the cell cytoplasm. It allows the staining of more than 95% of the viable cells, and avoids a loading saturation. It also provides sufficient stained cells for performing the calibration curve for each culture sample.

3. Incubate the cell suspension for 20 min at 37°C and agitate occasionally to prevent cell settling. The incubation time should be sufficiently long to allow cellular esterases to hydrolyze the ester form of the dye, but not too long, to minimize dye leakage out of the cell. Maximum fluorescence and ratio values were obtained after 10 min of incubation time and remained stable for 30 min. After longer incubation times, fluorescence intensity and ratio decrease.

4. Add 9 ml of EBSS buffer, remove ten aliquots of 10^6 cells, centrifuge them, and hold the cells on ice until analysis. Cell pellets can be conserved on ice for at least 2 h without any loss of fluorescence. Six aliquots are used for the calibration curve, two aliquots for a control, and two aliquots for the pH$_i$ measurement of the sampled cells.

5. Before analysis, suspend the pellets in 1 ml EBSS for the experimental samples or in high-[K$^+$] buffer, containing nigericin for the calibration samples.

B. Flow Cytometric Analysis

Flow cytometers used for intracellular pH measurement are usually equipped with an argon ion laser (488 nm, 200–400 mW of power). The emitted light is first separated from the fluorescence using a long-pass dichroic mirror at 500 nm. Then the fluorescence is separated into high and low wavelengths by a short-pass 550-nm dichroic mirror. These are further selected by a 525-nm band-pass filter (green fluorescence) and a 610-nm long-pass filter (red fluorescence).

Figure 3 shows an example of flow cytometric measurements obtained at the two wavelengths with hybridoma cells of high viability sampled after 25 h of a batch culture. For the two wavelengths, one observes a relatively large distribution of the number of cells as a function of the fluorescence intensity, which can be attributed to cellular differences in terms of cell volume, esterase content, dye uptake, or dye leakage, and possibly pH$_i$. When the cell count is represented as a function of the ratio of the green and red fluorescence, a narrower peak is obtained.

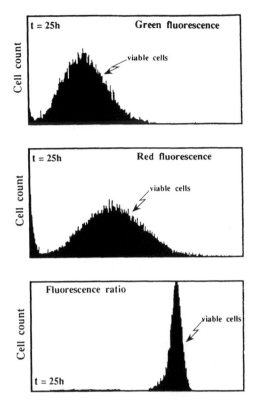

FIGURE 3 Distributions of green and red fluorescence, and of fluorescence ratio, obtained with hybridoma cells of high viability sampled after 25 h of a batch culture, stained with BCECF.

C. Calibration Curve Determinations

For each collected cell sample a calibration curve should be established to minimize possible interference related to the cells (volume, esterase content) and to the optical adjustments of the flow cytometer.

1. Take the previously prepared aliquots for the calibration, and suspend the cell pellets in 1 ml high-[K^+] buffers at different pH values (e.g., from 6.6 to 7.6). High-[K^+] buffers are obtained by mixing appropriate portions of two buffers, one containing 135 mM KH_2PO_4, 20 mM NaCl, and the other 110 mM K_2HPO_4, 20 mM NaCl, respectively.

2. Add 10 μM of nigericin from a stock solution of 1 mM in ethanol. Addition of nigericin equilibrates the intracellular pH of the stained cells to the pH of the surrounding buffer.
3. Incubate for 5 min on ice before analysis.
4. Measure the fluorescence intensity at the two selected wavelengths and calculate the fluorescence ratio between the green and red fluorescences.
5. The calibration curve is obtained by plotting the ratio as a function of the external pH. For the tested hybridoma cells, an increase of the the ratio with increasing pH was obtained (Fig. 4).

D. Cytometric Analysis in the Presence of Dead Cells

Flow cytometric measurement of pH_i for cells in culture may be complicated by the presence of substantial proportions of dead cells. When cells are grown in batch or continuous cultures, several phenomena, such as nutrient depletion, toxic metabolite accumulation, or shear stress, can markedly increase the cellular death rate. Dead cells are usually analyzed by the trypan blue method, which counts the cells unable to exclude the trypan dye because of their altered membrane permeability. Thus, dead cells may also be expected to yield different fluorescent responses when exposed to the pH_i indicators.

Figure 5 shows examples of histograms obtained for cells sampled after 120 h of a batch culture, when the viability is strongly reduced. At both the red

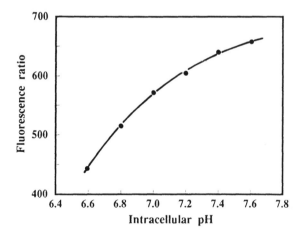

FIGURE 4 Calibration curve showing the relation between the fluorescence ratio and the intracellular pH.

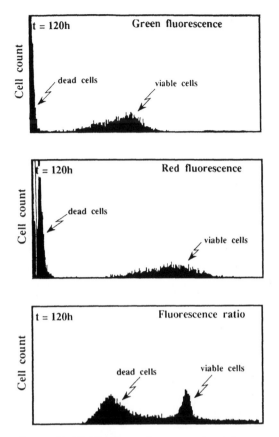

FIGURE 5 Distributions of green and red fluorescence, and of fluorescence ratio, obtained with hybridoma cells of low viability sampled after 120 h of a batch culture, stained with BCECF.

and green fluorescence wavelengths, one obtains a bimodal distribution of cells as a function of fluorescence intensity, with a much higher number of cells at the lower intensities. This part of the signal clearly corresponds to the dead cells, which are expected to have a lower level of esterases and a more rapid leakage of the fluorescent dye. This bimodal distribution is also observed for the cell count versus fluorescence ratio distribution. To measure only the pH_i of living cells, a gating procedure is applied that eliminates the contribution of the dead cells to the measured fluorescence ratios.

IV. INTRACELLULAR pH MEASUREMENTS DURING HYBRIDOMA CULTURES

The use of the protocol previously described for intracellular pH determination is illustrated for OKT3 hybridoma cells grown in batch and continuous 4-L bioreactors with controlled external pH at 7.0 and pO_2 at 50% of air saturation. The culture medium is basal DMEM medium, supplemented with 7% FCS, 25 mM glucose, 4.5 mM glutamine, 2% MEM essential amino acids, 1% MEM nonessential amino acids and 1% antibiotics.

A. Hybridoma Cultivated in a Batch Bioreactor

Figure 6 shows the typical kinetics of a batch culture of hybridoma cells culti-vated at a controlled medium pH of 7.0. The viable cell density increases

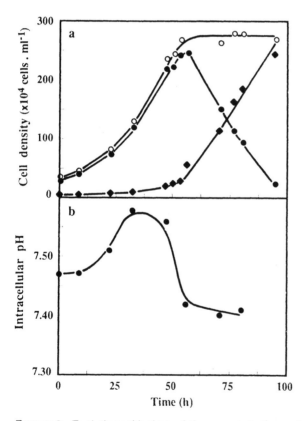

FIGURE 6 Evolution with time of the concentrations of viable (●), dead (♦), and total (○) cells, and of the intracellular pH, during a hybridoma batch culture.

from an initial value of 2.5×10^5/ml, to a maximum level of 2.5×10^6/ml after 55 h of culture. For longer culture times the concentration of viable cells rapidly decreases, mainly because of the depletion of essential nutrients, such as glucose, glutamine, or other amino acids.

Although the external pH is controlled at 7.0 during the culture, the measured intracellular pH shows important variations [12]. It first rises to a maximal value of 7.55 that is maintained during the growth phase. During the decline phase the pH_i of the remaining living cells gradually decreases to a value of 7.4.

B. Hybridoma Cultivated in a Continuous Bioreactor

In recent years, continuous cultures of mammalian cells have received increasing attention. First, they represent efficient production systems for cells and proteins, especially when they are operated in the perfusion mode. Second, cultivating cells in continuous, controlled bioreactors can provide original information on the kinetics, physiology, and metabolism of the cells.

When, after an initial batch phase, the culture is operated with a continuous feeding of nutrients and removal of products, one generally observes a transient phase during which the different concentrations change with time followed by a steady-state characterized by a time-constant concentration of nutrients, cells, and metabolites. Critical operational parameters for continuous cultures are the dilution rate, the feed medium composition, and the medium pH.

Figure 7 shows an example of a continuous culture of hybridoma cells performed with a standard feed medium and at a controlled pH of 7.0. The dilution rate, which is initially fixed at 0.01 h^{-1}, is increased stepwise when each steady-state is reached up to a final value of 0.04 h^{-1}. During the culture, which was maintained for 900 h, the viable cell density remains relatively constant near 2×10^6/ml, whatever the dilution rate. A markedly lower viability, however, is obtained at the lowest dilution rate.

When measuring the intracellular pH, important variations are again observed. Whereas, at inoculation, pH_i is near 7.2, it increases to 7.4 at the lowest dilution rate of 0.01 h^{-1}. It further rises to 7.6 when the dilution rate is increased to 0.03 and 0.04 h^{-1}. Thus, in both batch and continuous cultures, pH_i can be significantly different from the external pH and is strongly dependent on the culture conditions and the stage of the culture.

C. Relation Between pH_i and the Actual Specific Growth Rate

A possible explanation for the observed pH_i variation in the different culture systems is the relation that may exist between pH_i and the changing cell-specific growth rate. In batch culture the specific growth rate initially increases before it

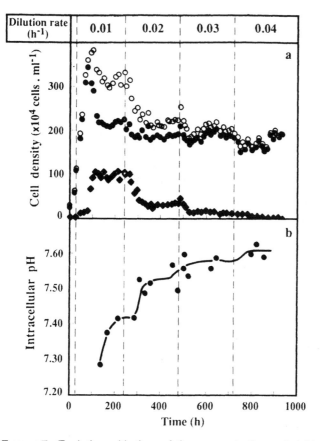

FIGURE 7 Evolution with time of the concentrations of viable (●), dead (♦), and total (○) cells, and of the intracellular pH, during a continuous hybridoma culture, performed at various dilution rates.

gradually decreases owing to nutrient depletion and metabolite accumulation. In continuous culture, on the other hand, the specific growth rate directly depends on the imposed dilution rate. When reaching steady state, for example, the specific growth rate is equal to the dilution rate. Increasing the dilution rate of continuous cultures thus enables the cells to be observed at increasing specific growth rate.

In Figure 8 the intracellular pH is plotted as a function of the specific growth rate for the batch and continuous hybridoma cultures previously shown. As seen, for both cultures, there is a general trend of increasing pH_i with increasing growth rate in the two culture modes. As observed for other cell lines,

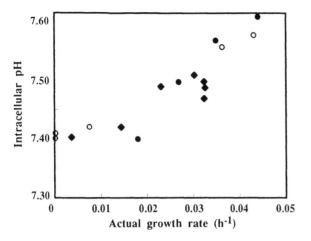

FIGURE 8 Relation between pH_i and the actual specific growth rate: (○) batch culture of Fig. 6, (●) continuous culture of Fig. 7, (◆) batch culture without pH control.

the intracellular pH of hybridoma seems to be closely related to the cellular growth state.

V. PERSPECTIVES

According to recent physiological investigations of mammalian cells grown in bioreactors, flow cytometry appears as a powerful technique to monitor the intracellular pH. It yields reliable measurements when using molecular probes that penetrate the cells and generate pH-sensitive dyes, provided care is taken in the treatment of the collected cells and in establishing a calibration curve for each sample. So far, most results have been reported with BCECF as a fluorescent probe. Improved results can be expected with a more recent class of dyes that have a greater pH-dependent shift in emission wavelength, such as SNARF (See Chap. 9).

Measurement of intracellular pH by flow cytometry undoubtedly offers new perspectives in the area of mammalian cell technology, both for better fundamental understanding of the regulation of cell metabolism and for improved operation of bioreactors. Further insights may be achieved by taking into consideration the large differences, up to 0.6 units for hybridoma cells, which may exist between the external and internal pH. Because of the high sensitivity of cellular processes to pH, this pH difference can have a marked effect on the regulation of cellular growth and metabolism.

A major finding with flow cytometry is that, at an externally controlled pH, the intracellular pH increases with the specific growth rate. This raises a basic question on the actual interdependency between pH_i and cellular growth: Is pH_i a result of the growth activity of the cell, or conversely, does pH_i, which is controlled by the cell metabolism and environment, directly determine the cellular growth rate?

The observed dependency of pH_i on the growth rate of cells in culture also provides a general framework for understanding the mechanism of action of substances that inhibit or stimulate the growth of cells. The often-reported inhibitory influence of excreted ammonia and lactate should be related to induced modifications in intracellular pH [6,7,10]. The kinetics of cell death, which are strongly dependent on the environmental composition, may also be analyzed in terms of dependency on the intracellular pH, especially when considering the respective contributions of the processes of apoptosis and necrosis [13–15].

Other possible effects of intracellular pH that should be investigated concern the processes of protein expression [16,17]. Does pH_i have a direct influence on the specific rate of protein synthesis and excretion and, consequently, on the relation between the specific rate of protein production and cell growth? Does pH_i interfere with the posttranslational modifications of the proteins and, for example, contribute to the macro- and microheterogeneity in protein glycosylation that has been reported under different culture conditions?

A future challenge for biochemical engineers is to integrate both the new qualitative and the quantitative understanding brought by the measurement of internal pH into intracellular models capable of simulating the kinetics of mammalian cells' metabolism in mass culture. A first single-cell model for CHO cells gives a direct linkage between intracellular pH control and cellular metabolism [18]. Among known mechanisms that regulate and disturb pH_i, the kinetics of the Na^+–H^+ antiport, the effect of lactate and ammonia, proton leakage and external pH, all are included in the model. The inhibition at low pH of protein synthesis, glycolysis, and nutrient uptake are also described.

Besides the expected progress in mastering cell physiology, the measurement of intracellular pH is also likely to bring novel approaches toward the operation of mammalian cell processes. For instance, for the production of proteins in fed-batch or perfusion reactors, following the pH_i may lead to a more efficient nutrient-feeding strategy [19] or an optimal external pH profile that maximizes cell growth in a first stage and protein synthesis in a second stage. Monitoring pH_i can also provide a way to detect possible dysfunctioning as well as the means for better control of cell cultures, thereby leading to improved process consistency.

In conclusion, given the development of flow cytometry to measure pH_i, together with the acquisition of new knowledge on the relation between the intracellular pH and cellular metabolism, the monitoring of pH_i, is likely to become a more widespread technique for the control and optimization of cell cultures for the production of cells, proteins, or viruses.

REFERENCES

1. G. Schmid, H. W. Blanch, and C. R. Wilke, Hybridoma growth, metabolism and product formation in Hepes-buffered medium: II. effect of pH, *Biotechnol. Bioeng. 12*:633 (1990).
2. W. M. Miller, H. W. Blanch, and C. R. Wilke, A kinetic analysis of hybridoma growth and metabolism in batch and continuous suspension culture: Effect of nutrient concentration, dilution rate and pH, *Biotechnol. Bioeng. 33*:947 (1988).
3. R. Nuccitelli and D. W. Deamer, *Intracellular pH: Its Measurement, Regulation and Utilisation in Cellular Functions*, A. R. Liss, New York, 1982.
4. E. Musgrove, M. Seaman, and D. Hedley, Relationship between cytoplasmic pH and proliferation during exponential growth and cellular quiescence, *Exp. Cell Res. 172*:65 (1987).
5. C. Van Haelst-Pisani, E. J. Cragoe, T. L. Rothstein, Cytoplasmic alkalinisation produced by the combination of anti-immunoglobulin antibody plus cytochalasin D in murine B lymphocytes, *Exp. Cell Res. 183*:251 (1989).
6. A. McQueen and J. E. Bailey, Growth inhibition of hybridoma cells by ammonium ion: Correlation with effects on intracellular pH, *Bioproc. Eng. 6*:49 (1991).
7. K. Martinelle and L. Häggström, Mechanisms of ammonia and ammonium ion toxicity in animal cells: Transport across cell membranes, *J. Biotechnol. 30*:339 (1993).
8. G. Valet, A. Raffael, L. Moroder, E. Wunsch, and G. Ruhenstroth-Bauer, Fast intracellular pH determination in single cells by flow cytometry, *Naturwissenschaften 68*:265 (1981).
9. M. Al-Rubeai and A. N. Emery, Flow cytometry in animal cell culture, *Biotechnology 11*:572 (1993).
10. S. S. Ozturk, M. R. Riley, and B. O. Palsson, Effects of ammonia and lactate on hybridoma growth, metabolism, and antibody production, *Biotechnol. Bioeng. 39*:418 (1992).
11. E. D. Wieder, H. Hang, and M. H. Fox, Measurement of intracellular pH using flow cytometry with carboxy-SNARF-1, *Cytometry 14*:916 (1993).
12. M. Cherlet, N. Petitpain, M. Dardenne, P. Franck, J. M. Engasser and A. Marc, Intracellular pH evolution during batch cultures using flow cytometry, *Animal Cell Technology: Products of Today, Prospects for Tomorrow* (R. E. Spier, J. B. Griffiths, and W. Berthold, eds.), Butterworth, London, 1994, p. 351.
13. S. Morana, J. Li, E. W. Springer, and A. Eastman, The inhibition of etoposide-induced apoptosis by zinc is associated with modulation of intracellular pH, *Int. J. Oncol. 5*:153 (1994).
14. S. Mercille and B. Massie, Induction of apoptosis in nutrient-deprived cultures of hybridoma and myeloma cells, *Biotechnol. Bioeng. 44*:1140 (1994).
15. R. P. Singh, M. Al-Rubeai, C. D. Gregory, and A. N. Emery, Cell death in bioreactors: A role for apoptosis, *Biotechnol. Bioeng. 44*:720 (1994).

16. C. F. Goochee and T. Monica, Environmental effects on protein glycosylation, *BioTechnology 8*:421 (1990).
17. M. C. Borys, D. I. H. Linzer, and E. T. Papoutsakis, Ammonia affects the glycosylation patterns of recombinant mouse placental lactogen-I by Chinese hamster ovary cells in a pH-dependent manner, *Biotechnol. Bioeng. 43*:505 (1994).
18. P. Wu, N. G. Ray, and M. L. Schuler, A computer model for intracellular pH regulation in Chinese hamster ovary cells, *Biotechnol. Prog. 9*:374 (1993).
19. G. K. Sureshkumar and R. Mutharasan, Intracellular pH-based controlled cultivation of yeast cells: II. Cultivation methodology, *Biotechnol. Bioeng. 42*:295 (1993).

9

The Relationship Between Intracellular pH and Cell Cycle in Cultured Animal Cells Using SNARF-1 Indicator

JONATHAN P. WELSH AND MOHAMED AL-RUBEAI
The University of Birmingham,
Birmingham, England

I. INTRODUCTION

The internal pH (pHi) of animal cells is an important physiological parameter, as cells can function only in a narrow range of pHi values (near neutral). The regulation of pHi is achieved through the action of the Na^+–H^+ ion exchanger [1,2]. Other pumps are also likely to play a role (e.g., the HCO_3^-–Cl^-exchanger). Cells have a large negative membrane potential, which would cause the passive diffusion of H^+ ions across the plasma membrane. Without the uphill extrusion of H^+ ions out of the cells through the action of the Na^+–H^+–exchanger, the cytoplasm of the cells would be too acidic for the cells to function. Through the action of the Na^+–K^+ ATPase, the concentration of Na^+ inside the cell is kept well below the concentration of Na^+ outside the cell. This provides the thermodynamic-driving force for H^+ extrusion. However, a greater efflux of H^+ can lead to an increase in pHi, a condition that is normally associated with an increase in cell activity.

163

Flow cytometry allows many of the changes during the culturing of animal cells, particularly those in which metabolic activity is related to the cell cycle phase, to be followed in detail [3, 4]. Cytoplasmic pH is an example of a parameter that can be used to monitor cell cultures and one that is amenable to measurement using flow cytometry [5; see also Chap. 8]. In monitoring hybridoma batch culture, for instance, the pHi increases after inoculation, peaks in the midexponential phase, and drops during the later phases. During hybridoma culture, the percentage of S-phase cells increases to a maximum during the exponential phase and drops back to a minimum at the stationary phase [6]. Therefore, it is possible that the increase in pHi is due to the increase in the number of S-phase cells.

We are interested in using cell cycle analysis to understand and predict the quality of the bioreaction process and to address the problem of modeling cell population dynamics. The possible association between cell cycle and pHi is particularly important in monitoring and control of continuous perfusion culture. An optimum pHi could be maintained by perfusion-rate regulation, so that the cells are kept in the actively dividing growth phase characterized by a high proportion of S- and G_2-phase cells.

In this chapter we demonstrate how flow cytometry can be used to show that this increase in pHi occurs during the stages of the cell cycle noted for increased cell activity; that is, the S and G_2 phases. This is done by (a) directly sorting cells into their respective phases and measuring pHi; (b) measuring the pHi of cells of different sizes from a population, assuming that those cells from the S and G_2 phases are larger than those from the G_1 phase of the cell cycle. Evidence to prove such an assumption is also provided. We describe how amenable flow cytometry is for making reliable pHi measurements using a pH-sensitive probe, carboxy-SNARF-1-AM.

II. MATERIALS AND METHODS

A. Carboxy-SNARF: Staining and Analysis

To make direct measurements of the pHi, cells were stained with the pH-sensitive probe carboxy-SNARF-1-AM (seminaphtorhodafluor-1-acetoxymethyl ester) and analyzed by flow cytometry. SNARF (available from Molecular Probes, Inc.) is derived from fluorescein. It enters the cell as a nonpolar ester that passes through the membrane passively, but once in the cytoplasm, it is hydrolyzed by esterases into a polar compound that cannot pass through membranes easily and so accumulates inside the cell. The main advantage of SNARF as a pHi indicator is the useful ratioing properties that it exhibits. It shows large changes in pHi-dependent fluorescence (Fig. 1). The acidic form shows an excitation maximum between 518 and 548 nm, and an emission peak

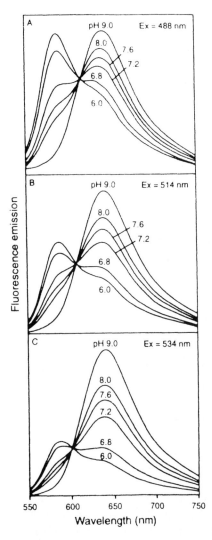

FIGURE 1 Fluorescence emission spectra of SNARF-1 as a function of pH at excitation wavelengths of (A) 488 nm, (B) 514 nm, and (C) 534 nm. (Courtesy of Molecular Probes Inc.)

at 587 nm. The basic form shows an excitation maximum at 574 nm and an emission maximum at 636 nm. When excited at 488 nm by an argon laser so that the fluorescence ratio of basic and acidic forms is measured, the ratio is seen to increase with increasing pHi. Measuring a fluorescence ratio allows variables, such as intracellular dye concentrations, probe leakage, and cell thickness, to be canceled out as the effects of these will be the same at each wavelength. Indeed, van Erp et al. [7] illustrated that pH measurements were independent of dye load. There have been several studies of the characteristics of SNARF that have established some of the idiosyncrasies of the stain [7–10]. Seksek et al. [8] showed that the fluorescence intensity of the probe dropped by 25% when the temperature shifted from 25 to 37°C. The same group found that the fluorescence intensity of the probe inside the cell reached a steady level after 20 min. They also showed significant probe leakage from the cell some 20 min after the end of the loading period. However, as flow cytometry measurements are completed within this time, it should not greatly affect results, as demonstrated by van Erp et al. However, it does show SNARF to be unsuitable for measuring the pHi of cells by conventional spectrofluorimetry.

Seksek et al. [10] also studied the effects of major cellular macromolecules on the fluorescence resulting from the probe. Their results indicated that SNARF is likely to interact with only intracellular proteins, but not with DNA or membrane phospholipids. About 20% of the dye is bound or trapped in cell organelles. However, the level of SNARF-1 stabilizes, probably corresponding to an equilibrium of the probe between target molecules, which emphasizes the requirement for calibration for each cell sample studied. No evidence for compartmentalization of the probe was found.

A calibration curve is needed to link the fluorescence ratio to pHi. This is accomplished by allowing cells stained with SNARF to equilibrate in high-concentration K^+ buffers of different pH values containing the H^+ ionophore nigericin, to allow the equilibration of internal and external H^+ and, hence, internal and external pH values. Nigericin does exhibit a degree of cytotoxicity; this is another reason to perform the flow cytometric analysis within 30 min after the addition of nigericin. A typical calibration curve is shown in Figure 2. The line is fitted through the points using a linear regression analysis. The pHi of the sample can be found from its fluorescence ratio by reading off such a calibration curve.

The procedure we used for loading hybridoma cells is based on methods already established [11]. Briefly, 6×10^6 cells per milliliter were spun down in a sterile 25-ml vial at 1000 rpm for 5 min and resuspended in 2.5 ml of HEPES-buffered medium. Carboxy-SNARF-1-acetomethyl ester (1 mM in DMSO) was added to give a final concentration of 10 μM. The cells were then incubated for 30 min at 37°C, with frequent mixing to ensure homogeneity of the sample. Calibration buffers in the pH range of 6.5–8.0 were prepared, and nigericin was added to them to a final concentration of 2 μg/ml. After incubation, the cells were spun down at 1000 rpm for 3 min. A pink pellet was observed. The supernatant was discarded and the pellet resuspended in 3 ml of fresh HEPES-buffered medium, split up into flow cytometry tubes, and spun down. The supernatant was carefully removed, and the cells resuspended in the range of buffers

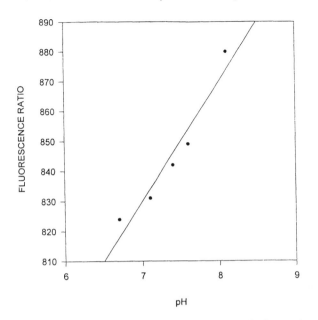

FIGURE 2 A typical calibration curve. The pHi of samples can be found using the linear relationship between fluorescence ratio and pH.

to create the calibration curve, or in fresh medium to make up the sample to be tested. The cells were allowed to equilibriate in the buffers for 5 min before being taken for analysis on the flow cytometer. Table 1 illustrates the high-K^+ calibration buffers used, and Table 2 shows the combinations of these buffers used to give the desired pH values.

All pHi analyses were performed on a Coulter EPICS Elite flow cytometer. The machine's argon laser, with an emission peak at 488 nm, is used to excite the dye, and the fluorescence at the red to yellow end of the spectrum is detected after filtering the emitted light. The emitted light first passes a 488-nm dichroic filter, which reflects 488-nm–laser light to provide a measure of side scatter. The remaining light from emission of the SNARF-1 passes through the 488-nm dichroic filter to a 600-nm dichroic filter, which reflects light of 600-nm. This light then encounters a 575-nm band-pass filter that lets light of only that wavelength pass to the photomultiplier tube detector. The light not reflected by the 600-nm dichroic filter is filtered by a 635-nm band-pass filter, which lets light at only this red end of the spectrum through to a detector. After conversion into a voltage proportional to its intensity, and after signal processing, the fluorescent emission gives rise to a number of useful measurements, as shown in Figure 3.

TABLE 1   Components Needed for High K+ Calibration Buffers

Buffer A	Buffer B
135 mM KH$_2$ PO$_4$	135 mM K$_2$H PO$_4$
20 mM NaCl	20 mM NaCl
1 mM MgCl	1 mM MgCl
1 mM CaCl$_2$	1 mM CaCl$_2$
10 mM glucose	10 mM glucose

TABLE 2   Combinations of Buffers[a] A and B to Give Range of pHi Values

pH	Volume of A (ml)	Volume of B (ml)
6.4	18.40	6.60
6.8	12.75	12.25
7.0	9.75	15.25
7.4	9.75	20.25
8.0	1.30	23.70

[a]The volume of each buffer is made up to 50 ml with distilled water, and nigericin added to 2 µg/ml to allow the buffers to permeate the cell membranes and equilibriate with the cells.

1. *Side scatter against forward scatter*: to distinguish live cells from dead cells and debris
2. *The fluorescence ratio* (635 nm against 575 nm)
3. *The cell count against fluorescence ratio*: gives an indication of the resolution of the staining and can be analyzed to give the mean fluorescence ratio for a particular population of cells

Each sample is analyzed by "gating" the live cell population as shown in Figure 3. The mean fluorescence ratio of the samples can be related to the internal pH found by direct reading off the calibration curve.

B. Cell Sorting

Sorting was carried out using a Becton-Dickinson FACS 440, fitted with an argon laser, with excitation at 440 nm. The SNARF-1-stained cells were sorted according to their fluorescence ratios. Two samples were collected directly into

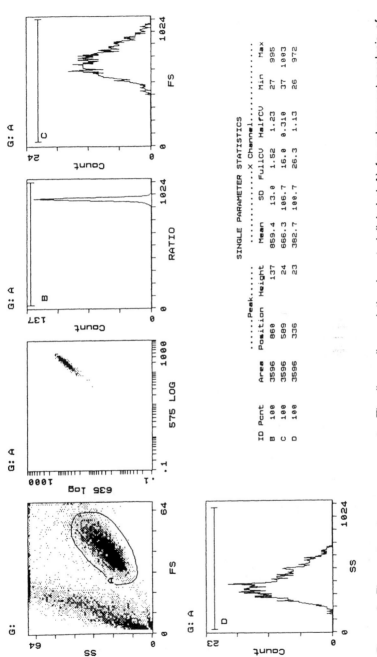

SINGLE PARAMETER STATISTICS

ID	Pcnt	Area	Position	Height	Mean	SD	FullCV	HalfCV	Min	Max
B	100	3596	860	137	859.4	13.0	1.52	1.23	27	995
C	100	3596	589	24	666.3	106.7	16.0	0.310	37	1003
D	100	3596	336	23	382.7	100.7	26.3	1.13	26	972

FIGURE 3 Flow cytometry measurements. The live cell population is gated (labeled A) for subsequent analysis of fluorescence ratio, forward scatter, and 90° scatter.

G: AB

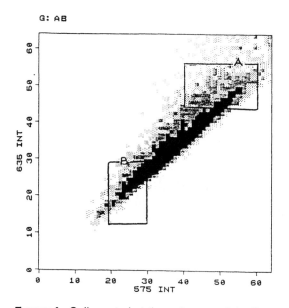

FIGURE 4 Cells sorted at the extremes of the fluorescence ratios. Boxes A and B indicate cells with high and low ratio, respectively.

70% ethanol, one of these subpopulations displayed a high fluorescence ratio (575/640 nm) the other displayed a low fluorescence ratio, as indicated in Figure 4. The two subpopulations of cells were then washed with phosphate-buffered saline (PBS), stained with 50 μg/ml propidium iodide, and analyzed for cell cycle distributions.

III. RESULTS

A. Direct Relationship Between Internal pH and Cell Cycle

The cell cycle distributions of the subpopulations are shown in Figure 5. Cells with a low fluorescence ratio were 81.8% G_1. Cells displaying the higher fluorescence ratio and, therefore, of higher pHi, were mainly from the S and G_2/M phases of the cell cycle. The pHi of the subpopulations was worked out by gating cells equilibrated in the range of buffers at the low and high ratios of fluorescence and constructing a calibration curve. The pHi of cells with a high fluorescence ratio was 7.6, whereas that of cells with a low ratio of fluorescence was 7.4.

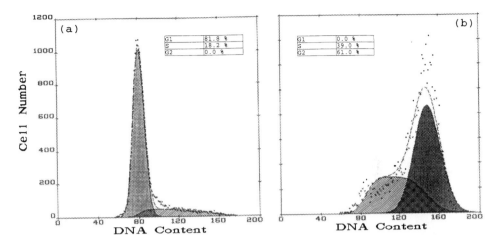

FIGURE 5 Cell cycle profiles of cells with low (a) and high (b) fluorescence ratio.

B. Indirect Relationship Between Internal pH and Cell Cycle

An indirect relationship is seen to exist between pHi and cell cycle by making the assumption that cells with a higher DNA content are from the G_2 phase, whereas cells with a lower DNA content are from the G_1 phase. The live cell population on the forward-scatter versus side-scatter dot plot can be gated into several subpopulations, according to their cell size (Fig. 6). Each gate is subsequently analyzed for fluorescence ratio and cell size. A typical result is shown in Figure 7, with small cells having a pHi of 7.2, whereas the larger cells have a pHi of 7.5. The cells with the largest forward scatter had the higher DNA content (i.e., from the G_2/M phases.

IV. DISCUSSION

The cell cycle phase is a fundamental physiological property indicating, for example, the potential of a population for growth and expansion. An intracellular alkalization has been linked to cell proliferation and synthetic activity. This is true for cells as diverse as yeasts, slime molds, ciliates, and lymphocytes. At higher internal pH, it has been noted that cells are stimulated from quiescence to activity. Ceccarini and Eagle [12,13] noted that contact inhibition of cultured fibroblasts was avoided if the external pH (pHe) was held at a level between 6.9 and 7.8. Chambard and Pouyssegur [19] showed the ribosomal S6 protein was

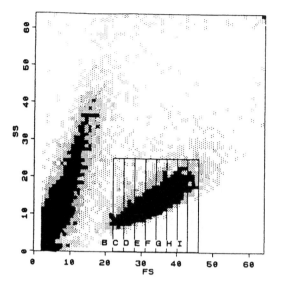

FIGURE 6 Gating of live cell population into eight subpopulations according to cell size. Each of these subpopulations was subsequently analyzed for fluorescence ratio.

more phosphorylated following intracellular alkalisation and this showed a correlation with the increase in the rate of protein synthesis.

Musgrove et al. [14] showed that acidification of the growth medium to a pH of 6.5 resulted in intracellular acidification and a marked increase in cells entering a quiescence state. The cells seemed to arrest at a particular point in the G_1 phase which, it was suggested, would be pHe-sensitive. Following alkalization of the medium, the cells reentered the growth cycle. However, it was not possible to characterize quiescent cells with a distinct lower internal pHi, indicating that cell cycling was reduced at the lower pHe values. These findings highlight the point that the change in pHi can regulate the entrance into the S phase of the cell cycle. Clearly, the pHi has a crucial role in the regulation of both DNA and protein synthesis and, thereby, progression through the cell cycle [15–22]. Lowering the pHe to slightly below the value needed for for reinitiation of DNA synthesis decreased protein synthesis and delayed the entry of cells from the G_0/G_1 to the S phase [19]. It could be that proteins needed for DNA synthesis are inhibited at this nonpermissive pHe.

Our results are in general agreement with the previous findings and support the assertion that the pHi is cell cycle-dependent. But the difference in pHi values between G_1 and G_2 cells cannot, by itself, explain the wide range of pHi values seen over the duration of batch culture. Clearly, as the pHi changes

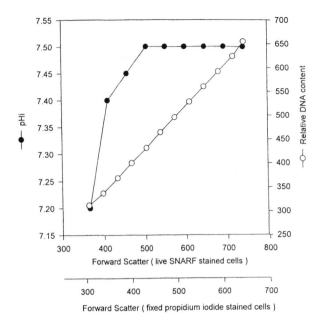

FIGURE 7 The indirect relationship between pHi and DNA as derived from their separate relationships with cell size.

during batch culture, with no pH control from 7.4 at the exponential phase to 6.8 at the decline phase, the external environment and specifically pHe have a greater influence on pHi. This influence is perhaps indirect owing to the deleterious effect on membrane integrity during the decline phase. Interestingly, the chemical and physical environment of the culture influences the progression of cells through the cell cycle by increasing the number of G_1 cells during the decline phase of batch culture [3,6]. Therefore, our conclusion, as demonstrated by this application of flow cytometry, is that cell cycle and pHi can be manipulated by the pHe and other environmental factors and that the influence of the cell cycle on the pHi is felt during mainly the growth phase of batch culture, but diminishes during the decline phase, when cells begin to die and their membranes become damaged.

REFERENCES

1. J. L. Seifter and P. S. Aronson, Properties and physiologic roles of the plasma membrane sodium-hydrogen exchanger, *J. Clin. Invest.* 78:859 (1986).
2. S. Grinstein and A. Rothstein, Mechanisms of regulation of the Na^+/H^+ exchanger, *J. Membr. Biol.* 90:1 (1986).

3. M. Al-Rubeai, S. Chalder, R. Bird, and A. N. Emery, Cell cycle, cell size and mito-chondrial activity of hybridoma cells during batch cultivation, *Cytotechnology 7:*179 (1991).
4. M. Al-Rubeai, M. Kloppinger, G. Fertig, A. N. Emery, and H. G. Miltenburger, Monitoring of biosynthetic and metabolic activity in animal cell culture using flow cytometric methods, *Animal Cell Technology: Developments, Processes and Products* (R. E. Spier, J. B. Griffiths, and C. MacDonald, eds.), Butterworth-Heinemann, Oxford, 1992, p. 301.
5. M. Cherlet, N. Petitpain, M. Dardenne, P. Franck, J. M. Engasser, and A. Marc, Intracellular pH evolution during batch cultures using flow cytometry, *Animal Cell Technology: Products of Today, Prospects for Tomorrow* (R. E. Spier, J. B. Griffiths, and W. Berthold, eds.), Butterworth-Heinemann, Oxford, 1994, p. 351.
6. M. Al-Rubeai, A. N. Emery, and S. Chalder, *Flow cytometric study of cultured mammalian cells, J. Biotechnol. 19:*67 (1991).
7. P. E. J. van Erp, M. J. J. M. Jansen, G. J. de Jongh, J. B. M. Boezeman, and J. Schalkwijk, Ratiometric measurement of intracellular pH in cultured human keratinocytes using carboxy-SNARF-1 and flow cytometry, *Cytometry 12:*127 (1991).
8. K. J. Buckler and R. D. Vaughan-Jones, Application of a new pH-sensitive fluoro-probe (carboxy-SNARF-1) for intracellular pH measurement in small, isolated cells, *Pflugers Arch. 417:*234 (1990).
9. S. Bassnett, L. Reinisch, and D. C. Beebe, Intracellular pH measurement using single excitation–dual emission fluorescence ratios, *Am. J. Physiol. 258:*C171 (1990).
10. O. Seksek, N. Henry-Toulme, F. Sureau, and J. Bolard, SNARF-1 as an intracellular pH indicator in laser microspectrofluorometry: a critical assessment, *Anal. Biochem. 193:*49 (1991).
11. P. S. Rabinovitch and C. I. June, Intracellular ionized calcium, membrane potential, and pH, *Flow Cytometry: A Practical Approach* (M. G. Omerod, ed.), Oxford University Press, Oxford, 1990, p. 179.
12. C. Ceccarini and H. Eagle, Induction and reversal of contact inhibition of growth by pH modification, *Nature 233:*271 (1971).
13. C. Ceccarini and H. Eagle, pH as a determinant of cellular growth and contact inhi-bition, *Proc. Natl. Acad. Sci. USA 68:*229 (1971).
14. E. Musgrove, M. Seaman, and D. Hedley, Relationship between cytoplasmic pH and proliferation during exponential growth and cellular quiescence, *Exp. Cell Res. 172:*65 (1987).
15. S. S. Ober and A. B. Pardee, Intracellular pH is increased after transformation of Chinese hamster embryo fibroblasts, *Proc. Natl. Acad. Sci. USA 84:*2766, (1987).
16. D. F. Gerson, The relation between intracellular pH and DNA synthesis rate in proliferating lymphocytes. *Intracellular pH: Its Measurement, Regulation and Utilization in Cellular Functions* (R. Nuccitelli and D. W. Deamer, eds.), Alan R. Liss, New York, 1982, p. 375.
17. P. J. Nielson, R. Duncan, and E. H. McConkey, Phosphorylation of ribosomal protein S6. Relationship to protein synthesis in HeLa cells, *Eur. J. Biochem. 120:*523 (1981).

18. G. Thomas, M. Siegmann, A. M. Kubler, J. Gordon, and J. de Asua, Regulation of 40S ribosomal protein S6 phosphorylation in Swiss mouse 3T3 cells, *Cell 19:*1015 (1980).

19. J.-C. Chambard and J. Pouyssegur, Intracellular pH controls growth-factor induced ribosomal protein S6 phosphorylation and protein synthesis in the G0–G1 transition of fibroblasts, *Exp. Cell Res. 164:*282 (1986).

20. G. Thomas, J. Martin-Perez, M. Siegmann, and A. M. Otto, Effect of serum, EGF, $PGF_{2\alpha}$, and insulin on S6 phosphorylation and initiation of protein and DNA synthesis, *Cell 30:*235 (1982).

21. P. W. Rossow, V. G. H. Riddle, and A. B. Pardee, Multiple phosphorylation of ribosomal protein S6 during the transition of quiescent 3T3 cells into early G1 and cellular compartmentalization of the phosphate donor, *Proc. Natl. Acad. Sci. USA 76:*4446 (1979).

22. J. Pouyssegur, A. Franchi, and P. Silvestre, Relationship between increased aerobic glycolysis and DNA synthesis initiation studied using glycolytic mutant fibroblasts, *Nature 287:*445 (1980).

10

Assessment of Cell Viability in Mammalian Cell Lines

XAVIER RONOT AND SYLVAIN PAILLASSON
Ecole Pratique des Hautes Etudes,
Institut Albert Bonniot, Grenoble, France

KATHARINE A. MUIRHEAD
Zynaxis Inc., Malvern, Pennsylvania

I. THE USEFULNESS OF CELL VIABILITY ASSESSMENT

There are many situations in which it may be important to evaluate the viability of cultured mammalian cells: for example, to optimize conditions for isolation (sorting, cloning) or cryopreservation; to aid in selection of optimal conditions and quality control for long-term cultures; to assess the effect of exposure to xenobiotics, ranging from drugs to biomaterials; and others. Flow cytometry is a particularly useful tool for monitoring the health of cultured cells because it can be used to measure a wide range of cell functions. However, this very flexibility can lead to confusion about exactly what is meant by "cell viability" or "cell death," as discussed later [1].

On occasion, it may be sufficient simply to identify *dead* cells as those incapable of excluding certain dyes and then to eliminate them from consideration for sorting or analysis. For example, cells of which membranes are no longer intact often adsorb antibodies nonspecifically, leading to false-positive

177

results when immunological staining is used to identify cells producing a certain type or amount of antigen. However, it may be desirable to monitor early changes in function that follow toxic events (e.g., decreased enzyme or metabolic activity, loss of specialized functions such phagocytosis, or membrane perturbations). Similarly, it is important to monitor recovery from initial damage to determine whether cells become static, eventually begin to grow again, or eventually lose membrane integrity and die. Ideally, one would like to be able to choose from a range of "viability" assays indicative of a wide variety of cell injuries [2].

II. THE CRITERIA FOR ASSESSING CELL VIABILITY

Despite some disagreement about the definition of the living, viable, or "intact" state, cells are usually characterized by various criteria related to viability assessment:

1. Size, shape, and refringency, which are well defined and conditioned by the intracellular organization and the level of cell hydration
2. The capacity to incorporate or concentrate dyes in certain organelles (lysosomes, mitochondria, nucleus) and to modify these dyes
3. The ability to exclude the influx of dyes, in relation to selective permeability The injured cell will have lost one or several of these characteristics; but, usually, when studied under the microscope, the cells will be analyzed for only one of these criteria.

In addition to these criteria, cell proliferation is also a parameter of interest, and clonogenicity is often measured when cell viability is concerned. However, this assay cannot provide an instantaneous estimate of the fraction of intact cells, the parameter of interest when studying the effects of xenobiotics or metabolic activities requiring cell integrity. Cell viability determination (i.e., the fraction of living cells in a population) is directly related to morphological, physicochemical, or biological differences (e.g., cell behavior relative to the accumulation of fluorescent dyes specific to cell components or compartments).

The use of fluorescent dyes is a widely applied means of investigation, and flow cytometry has improved cell viability assessment on the basis of the reliability, reproducibility, speed of analysis, and reduction of the subjective bias induced by microscopic observation of cells; hence, a large number of cells may be analyzed. However, it is imperative to know accurately the properties of fluorescent dyes, including their effects on cell proliferation and metabolism, as well as their physicochemical and spectral properties, especially when combining viability probes with other markers.

III. DIFFERENT METHODS OF ASSESSING CELL VIABILITY BY FLOW CYTOMETRY

When considering the available dyes, the main categories for assessing cell viability are the following:

Exclusion dyes
Vital dyes
Combinations of the two

For the results obtained, the main possibilities for assessing cell viability are the following:

Dead cell detection
Live cell detection
Combined live and dead cell detection

If we consider the flow cytometric capabilities, the main methods for assessing cell viability are the following:

Light absorption measurement
Light-scattering measurement
Fluorescence emission measurement, following staining with fluorescent dyes

These parameters provide information on modification in cell internal structure, loss in membrane integrity, and loss of metabolic activity.

A. Flow Cytometric Analysis of Unfixed Samples

1. The Light Absorption Approach

The magnitude of the extinction signal is dependent on the geometric cross-sectional area of the shadow of cells, reflecting the size of these cells, and also the scattering signal used for sizing of various type of cells. A linear correlation between the geometric cross section and the magnitude of the extinction signal was found when normal viable cells were used instead of particles in flow cytometric analysis [3].

Light absorption or transmittance measurements on entire cells pose technical problems that are not shared by fluorescence and light-scattering measurements. One such problem is that, usually, cells (even when stained) do not absorb more than a small fraction of the light passing through them. Therefore, fluctuations in the intensity of the incident beam have larger effects on transmittance measurements than on measurements of scattered light and fluorescence.

Moreover, transmittance measurements require relatively high optical resolution, as compared with scatter and fluorescence measurements. Lenses with high numerical apertures must be used to ensure that most of the light passing through the object under study is collected. For maximum accuracy in transmittance measurements, it is also necessary to match the refractive indices of the object under study and the medium in which it is suspended, to eliminate apparent absorption caused by the scattering of incident light out of the collection angle of the objective. In studies with intact cells, it is not generally possible to avoid such scattering by internal cellular structures (e.g., cytoplasmic granules), even when the refractive index of sheath, sample, and cell membranes are matched. Because of these limitations, light scattering is more commonly used than light absorption in commercial flow cytometers.

2. The Light-Scattering Approach

The light-scattering process is a collective response to the incident light, which consists in three components: reflection, refraction, and diffraction. Forward (small-angle) light-scattering measurement is used to obtain an indication of cell size. Right-angle light scattering is more sensitive to internal structural differences and to refractive index than is forward-angle light scattering.

Among cellular properties, other than the size, that influence the distribution of the scattered light are the differences in the refractive index between cells and the suspending medium, the internal structure of the cell, and the presence within or upon cells of material that absorbs strongly at the illumination wavelength used. If the refractive index of the cells were the same as that of the suspending medium, the cells would not scatter incident light. The difference in index between cells and the medium is maintained, at least partly, by the action of the membrane as a permeability barrier to water and solutes. Cells with damaged membranes (i.e., those cells that are identified as dead by uptake of exclusion dyes, such as trypan blue) have a lower refractive index and, thus, produce smaller forward-scatter signals. Such discrimination is less than perfect, especially when the sample contains cells of different types or size [4]. In addition, forward-angle, light-scattering intensity is strongly affected by the wavelength of light used, and UV excitation gives different scattering patterns than visible excitation [5]. Moreover, correlation between scattered light intensity and physical and biological properties of cells are affected by the precise range of angles over which light is collected [6]; for example, the shift of light-scattering values for intensely fluorescently stained cells in comparison with nonfluorescent ones. The combination of forward-angle light scattering and right-angle light scattering is more reliable than each parameter separately in discriminating between cells of heterogeneous sizes and shapes; it has become the most prevalent use of light scattering to analyze living and dead cell populations [4]. This

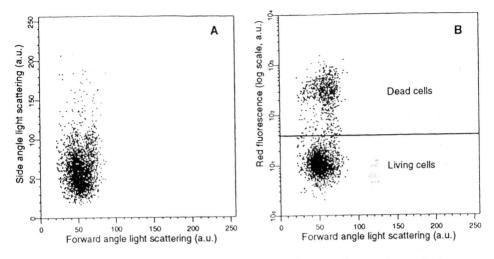

FIGURE 1 Determination of cell viability using light-scattering or dye-exclusion assay. (A) Analysis of K562 cells in culture using dual light scattering (forward light scattering vs. side light scattering. Twenty thousand cells were analyzed through a FACScan cytometer (Becton-Dickinson). PROCYT Loginserm software was used for data processing. Note that light scatter does not provide a clear distinction between living and dead cells. (B) Analysis of living and dead K562 cells using propidium iodide exclusion assay. Dead cells were discriminated by measuring forward-angle light scattering vs. fluorescence. Twenty thousand cells were analyzed through a FACScan cytometer (Becton Dickinson). PROCYT Loginserm software was used for data processing. (This work was carried out in collaboration with Drs. Mireille Favre and Marie-Christine Jacob, Laboratoire d'Immuno-Cytologie, Centre de Transfusion Sanguine, Grenoble, France.)

combination is routinely used to distinguish between unstained human leukocytes and the fraction of neutrophils, lymphocytes, and monocytes (Fig. 1).

3. The "Fluorescent" Approach

The availability of flow cytometers has stimulated interest in the development and application of fluorescent dyes that permit analysis and isolation of viable cells, with selected characteristics for either biochemical studies, or for further short- or long-term culture. Dyes used in such studies should, ideally, neither perturb the parameters measured nor compromise the viability of the cells being observed. Many of the dyes now in use fall short of this ideal when staining conditions do not induce cell perturbations. The conventional fluorescent dyes

TABLE 1 Major Fluorescent Probes for Assessment of Cell Viability

Name	MW	Excitation Emission (nm)	Specificity	Type	Refs.
Ethidium bromide	394	510/595 (bound to DNA)	DNA	Exclusion	11
Dihydroethidium	315	526/605 (oxydized to ethidium, which binds to DNA)	DNA	Vital	14
Propidium iodide	668	550/610 (bound to DNA)	DNA	Exclusion	17
Fluorescein diacetate	416	500/525 (converted to fluorescein)	Esterases	Vital	16
Rhodamine 123	381	500/545 (bound to mitochondria)	Mitochondria	Vital	18
Ethidium homodimer-1	857	528/617 (bound to DNA)	DNA	Exclusion	13
Hoechst 33342	562	340/450 (bound to DNA)	DNA	Vital	26,27

used may be characterized by their absorption and fluorescence emission wavelengths and intensities, and by whether they act as vital dyes or exclusion dyes (Table 1).

Exclusion Fluorescent Dyes

Dye exclusion methods are a means of quantifying cell death, as measured by loss of membrane integrity. A classic example of an exclusion dye, albeit a nonfluorescent one, is trypan blue, a tetrazoic dye that is excluded by living cells, but shows a diffuse incorporation when membrane integrity is lost [7]. It is frequently used in routine cytology, but it has several drawbacks. First, its high affinity for proteins leads to reduced concentrations in culture medium containing serum and, therefore, a reduced ability to identify nonviable cells [8]. Second, it also exhibits significant cytotoxicity and can lead to lysis of labeled cells, with subsequent underestimation of the fraction of nonviable cells. Finally, since viable cells remain unstained, trypan blue has little ability to detect early evidence of cell damage and gives poor correlation with assays based on cell growth or colony formation [9].

Erythrosin B, propidium iodide, and ethidium bromide are fluorescent dyes that, because of their charge, are excluded by cells with intact membranes. Erythrosin B is an acidic dye that is excluded by living cells [10], and its staining properties are not affected by protein concentration. Ethidium bromide and propidium iodide are phenanthridium derivatives, with a single- or double-positive charge, respectively. They selectively bind to double-stranded nucleic acids (both DNA and RNA), and this binding enhances their fluorescence efficiency, leading to bright red fluorescence, even in the presence of free dye [11] (see Appendix A). Since it has only a single positive charge, ethidium bromide is not as efficiently excluded by intact membranes, and dim staining of viable cells has been reported for some cell types [12], although dead cells can be distinguished by their much brighter staining. Ethidium homodimer has been described as being able to bind to nucleic acids 1000 times more tightly than ethidium bromide and to undergo a 40-fold increase in fluorescence on binding [13].

Vital Fluorescent Dyes

Vital dyes selectively accumulate in metabolically active cells by processes that occur in these living cells that would be altered or would stop in dead cells. Vital fluorescent dyes that are sensitive to changes in vital cell metabolism can be used to quantitate the toxic action of xenobiotics and may overcome some of the difficulties associated with dye exclusion tests. Moreover, the accumulation of color or fluorescence in intact cells can be used to provide an indication of the presence and relative activity of various enzymes in those cells.

The derivative Hoechst 33342 has been extensively used in flow cytometry to analyze and sort living cells according to their DNA content. Hoechst 33342 has similar properties to Hoechst 33258, but shows a greater membrane permeability, permitting the efficient staining of the nuclear DNA of living cells, without prior perturbation of the cell membrane by detergents or organic solvents. This probe stains the DNA of both living and dead cells and, as a consequence, is not the dye of choice to estimate cell viability or death when used alone. However, it does permit sorting of viable cells by DNA content, when combined with ethidium bromide or propidium iodide for dead cell exclusion.

Dihydroethydium (also called hydroethidine) is a fluorogenic substrate that monitors the oxidative capability of cells. This compound is a chemically reduced ethidium that may be dehydrogenated to ethidium after entering living cells. Ethidium is a known DNA intercalator and displays a red fluorescence when bound to nucleic acids. Because of the binding mode of ethidium, hydroethidium does not really act as a vital dye, since cells must be metabolically active to modify it to ethidium [14].

Rhodamine 123, a cyaninelike fluorescent dye, can enter living cells directly—without passage through endocytic vesicles and lysosomes—and is selectively accumulated by mitochondria. If no inhibitor or inhibitorlike molecules are added to cells and they are examined while in equilibrium with the dye, differences in fluorescence will be observed between damaged cells with intact membranes and those with normal mitochondrial function. Cells with deenergized mitochondria will take up less dye than cells with an intact energy metabolism; cells with cytoplasmic membrane damage sufficient to diminish or abolish the cytoplasmic membrane potential gradient will also take up less dye. Thus, equilibrium dye fluorescence measurements in cells can provide an indication of cell viability that is based on metabolic integrity as well as membrane integrity [15].

Fluorescein diacetate is a colorless, uncharged fluorogenic substrate that is enzymatically converted by living cells to fluorescein [16], a charged species that is retained by cells with intact membranes. Therefore, metabolically active cells accumulate fluorescein and become intensely green, although the fluorescein slowly diffuses into surrounding cells if analysis is not completed quickly. Therefore, more highly charged or less pH-sensitive analogues of fluorescein, such as BCECF, have become preferred for viability measurement [17].

Association of Fluorescent Dyes

Combinations of exclusion and vital dyes, such as fluorescein diacetate with erythrosin B or rhodamine 123 with ethidium bromide, as well as fluorescein diacetate with ethidium and rhodamine 123 with propidium iodide have been developed to obtain screening systems for assessment of membrane integrity, mitochondrial function, and viability [18,19] (see Appendix B). One of the major problems that arises with a dual fluorochrome system that is detected simultaneously is that spectral overlapping results because both green and red peak emissions are relatively close. Even with the use of interference band-pass filters, some spectral overlap will occur, eventually requiring fluorescence compensation.

The monitoring of both fluorescein diacetate and propidium iodide uptake may be useful as a general test for cell viability, and may also have distinct advantages over the use of either dye alone. Fluorescein diacetate is more sensitive to cell injury than conventional exclusion dyes, but it must be used in combination with flow cytometry to measure the differential dye uptake into viable and dead cells. Propidium iodide alone stains only nonviable cells, and thus gives no information concerning the metabolic integrity of the remaining cells.

Other combinations also allow the analysis of various cell physiological characteristics for a large population of cells and their correlation with cell viability. An example is the evaluation of acrosomal status of spermatozoa according to procedures that use monoclonal antibodies that react with the inner acrosomal membrane of human spermatozoa, and a supravital dye to estimate sperm viability and acrosomal status simultaneously (Fig. 2) [20].

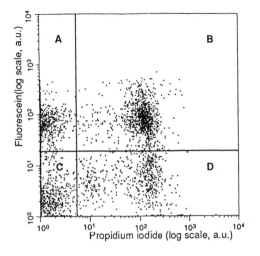

FIGURE 2 Simultaneous assessment of the acrosomal status and viability of human spermatozoa. The acrosomal reaction of mammalian spermatozoa, a modified form of exocytosis, is an essential event in fertilization. Spermatozoa cultured in BWW medium supplemented with human serum albumin (HSA) (Sigma) were treated with calcium ionophore A 23187 (Calbiochem) to induce an acrosomal reaction, labeled with monoclonal antibody GB24 (Theramex), then counterstained with propidium iodide (0.1 µg/ml, Sigma). Twenty thousand cells were analyzed through a FACScan cytometer (Becton Dickinson). Living spermatozoa are located in quadrants A (positive acrosomal reaction, 40%) and C (negative acrosomal reaction, 23%); dead spermatozoa are located in quadrants B (with positive acrosomal reaction, 28%) and D (negative acrosomal reaction, 9%). PROCYT Loginserm software was used for data processing. (This work was carried out in collaboration with Dr. Jean Cozzi, Laboratoire de Cytogénétique et Biologie de la Reproduction, Institut Albert Bonniot, Grenoble, France.)

B. Flow Cytometric Analysis of Fixed Samples

There are a number of cases in which the ability to distinguish between live and dead cells in fixed samples would be helpful. For example, it may be preferable to run fixed samples if biohazards are potentially present (as in cultured human cells), if access to the flow cytometer is limited (shared facility, instrument problems, and such), or if experimental protocols make immediate analysis difficult (late-day sample arrival, many courses with multiple sampling points, and so on). However, most fixation methods permeabilize cell membranes, whereas most of the methods for monitoring cell viability define *live cells* as those with membranes that are intact enough either to exclude certain dyes (e.g., propidium iodide or ethidium bromide), or to trap others (e.g., fluorescein diacetate or BCECF). In addition, most of the exclusion dyes are not covalently bound to their intracellular targets, but will redistribute over all cells in the population if cells are stained for viability and then fixed. This problem is compounded when immunofluorescence staining is required as part of the analysis, since nonspecific staining is often markedly enhanced in dead cells.

Although no general solution to this problem has yet been described, several alternatives exist that may assist cell culturists in selecting one appropriate to their particular needs. For example, treatment with DNase before fixation allows discrimination between cells that were intact (live) and those that were permeable (dead) at the time of exposure to the enzyme, since the DNA-staining intensity of dead cells is markedly reduced after such treatment. This approach has been used to study antigen-specific proliferative responses in cultured lymphocyte populations [21]. Alternatively, identification of dead cells in mixed leukocyte populations has been reported using prefixation exposure to ethidium monoazide, a photoactivatable analogue of ethidium bromide, followed by brief exposure to visible light to covalently label the dead cells, thereby avoiding redistribution of the label after fixation [22]; one limitation to the use of ethidium monoazide is that, as with ethidium bromide [13], low levels of staining have been reported in some types of viable cells [23]. Therefore, the ability of live cells to exclude either ethidium bromide or ethidium monoazide should be verified for the cell type of interest when adapting this method to other cell types.

A method for postfixation staining of mixed leukocyte populations with the laser dye LDS-751 (excitation wavelength: 488 nm, emission wavelength: 670 nm) has also been described [24]. This method has the advantage of being more compatible with immunofluorescence staining using fluorescein- and phycoerythrin-labeled antibodies; the bright LDS-751-labeled cells exhibit the decreased light-scattering characteristics that would be expected for dead cells. However, the mechanism of discrimination is undefined (perhaps relating to altered chromatin structure in damaged cells), and a surprisingly large proportion of leukocytes (> 60%) were reported as "damaged" by this method after standard whole blood lysis. Therefore, some independent measure of viability should be used to determine whether LDS-751 staining gives an equivalent answer when adapting this technique for use with other cell types.

Finally, a personal communication from Dr. Stan Berberich to one of the authors (K. A. Muirhead) suggested still another solution to the problem of retrospective viability determination. If fixation conditions can be modified to maintain the permeability differences, rather than causing all cells to become permeable, it may be possible to use a standard exclusion dye, such as propidium iodide, after fixation in exactly the same way that it is used for unfixed samples. As shown in Figure 3, fixation with a reduced concentration of formaldehyde or paraformaldehyde (0.1% instead of 1%) appears to have the desired effect for

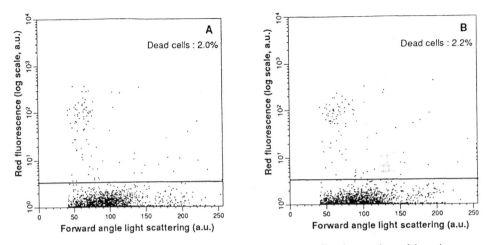

FIGURE 3 Dead cell fraction estimation on fresh and fixed samples of lymphocytes: Lymphocytes were obtained from human blood and analyzed after lysis by FACS lysing solution (Becton Dickinson). Aliquots of the sample were either (A) stained with propidium iodide (Sigma), 1 µg/ml; or (B) fixed with paraformaldehyde (Merck), 0.1% for 30 min, then stained with propidium iodide, 1 µg/ml. Of each aliquot, 20,000 cells were analyzed through a FACScan cytometer (Becton Dickinson) Both forward-angle light scattering (abscissa) and fluorescence (ordinate) were recorded. PROCYT Loginserm software was used for data processing. The upper part of the dot plot contains the dead cells, whereas the lower part contains the living fraction.

lymphocytes. The fraction of propidium iodide-positive cells is essentially identical whether the propidium iodide is added to an unfixed cell population or to the same population after "light" fixation. Although one might wonder whether the lower concentration of fixation is adequate to minimize biohazards, such as viral contamination, 0.1% paraformaldehyde fixation (used by Coulter automated Q-Prep system) has been demonstrated to reduce human immunodeficiency virus (HIV) infectivity to a degree similar to that found for commercial preparations containing higher percentages of fixatives [25]. However, variables such as the exact concentration of fixative, the amount of protein present, and so forth, may need to be reoptimized in adapting this method for cell culture.

IV. SOME PITFALLS

Important points to be discussed include the spectral properties of the fluorescent dyes, their potential cytotoxicity, and the effect of the enzymes used to detach anchorage-dependent cultured cells.

In multiple-labeling assays to determine cell viability and death, the absorption and fluorescence spectral ranges and maximum must be accurately known, and maximization of the spectral separation of the respective emissions is a basic step to avoid overlapping fluorescence spectra. In addition, the fluorescence spectra and quantum yields of some fluorescent dyes, particularly fluorescein, are highly environment-dependent (medium, pH, or other).

Cell viability assessment first requires determination of the effect of dye concentration on cell viability detection and on cell metabolism. For viability assessment, the literature describes propidium iodide concentrations ranging from 0.1 to 10 µg/ml. Hoechst 33342 is an example of a fluorescent dye that enters living cells without affecting the viability of the cells. This probe appears to inhibit chromosomal condensation and, thereby, normal mitosis, and induces a nonreversible G_2 accumulation, even at low concentrations (4 µM). Nonetheless, since Hoechst 33342 can traverse the cell membrane, the probe can also be removed from the chromatin, to a large extent, so that many cell lines retain full viability. Hoechst 33342 has been shown to enter cells by the same mechanism as colchicine (i.e., unmediated diffusion) [26,27]. There is no universal concentration available for all cell types, and the right concentration must be tested before assessing cell viability with a new fluorescent dye or a new cell type.

Flow cytometric assessment of cell viability requires carefully controlled use of cell disaggregation techniques, since enzymes, such as trypsin, have been reported to induce cell damage that can lead to artificially high estimates of cell death owing to membrane alteration or permeabilization [28].

V. CONCLUDING REMARKS

It appears that we have reached a point at which systematic application of multiple fluorochromes that display a selectivity for one or more classes of macromolecules or specific organelles is a practical reality. The technological developments that have assisted this research include the development of convenient culture as well as flow cytometric techniques.

Flow cytometry provides a method by which certain viability assays and cytotoxicity processes can be followed in terms of breakdown in cell physiological activities. It offers numerous advantages for study of cell viability (the great number of events analyzed, the suppression of subjective bias, and the multiparametric analysis), providing more information than would be obtained by conventional microscopic exclusion tests. However, the goal should always be to choose the simplest method and smallest number of fluorochromes necessary to identify live (or dead) cells. This simplifies the analysis and leaves most parameters free for assessment of other cell properties of interest.

APPENDIX

A. Dead Cell Assessment Using Propidium Iodide Exclusion Assay

Procedure for anchorage-dependent cells   (for 10^6 cells cultured in 25-cm^2 flasks):

1. Remove culture medium.
2. Wash the monolayer with PBS (Ca- and Mg-free).
3. Harvest the cells with 1 ml of trypsin/EDTA (0.1% w/v:0.02% w/v) solution in PBS.
4. Resuspend in 1 ml PBS or culture medium containing a trypsin inhibitor (soybean trypsin inhibitor or bovine serum) and monodisperse.
5. Add propidium iodide at a final concentration of 1 μg/ml before analysis.
6. Analyze the cell suspension using an argon ion laser tuned to 488 nm and measure the propidium iodide (red) fluorescence through a 630-nm wavelength filter.

Dead cells exhibit a bright propidium fluorescence, whereas living cells are colorless (see Fig. 1B).

B. Living or Dead Cell Assessment Using Combined Staining with Rhodamine 123 and Propidium Iodide

Material

1. Phosphate saline buffer (PBS) or other culture medium (phenol red free)
2. Stock solution of rhodamine 123, 1 mg/ml in distilled water, maintained at 4°C in the dark (good for 1–2 weeks)
3. Stock solution of propidium iodide, 1 mg/ml in distilled water, maintained at 4°C in the dark

Procedure for anchorage-dependent cells   (for 10^6 cells cultured in 25-cm^2 flasks)

Cell staining is directly performed on cell monolayers, since suspensions of adherent cells often evolve in cell aggregate.

1. Remote culture medium.
2. Add 5 ml rhodamine 123-containing medium (100 ng/ml in culture medium).

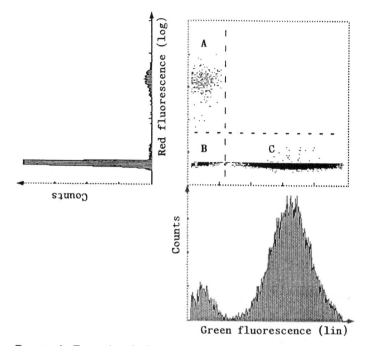

FIGURE 4 Example of discrimination of living or dead cells using dual fluo-rochrome assay (rhodamine 123 and propidium iodide). The dot plot provides three cell populations: (A) dead cells, rhodamine 123-negative and propidium iodide-positive; (B) cells with altered mitochondria, rhodamine 123- and propidium iodide-negative; and (C) living cells with energized mitochondria, rhodamine 123-positive and propidium iodide-negative. Projections as histograms are shown for each fluorescence. For each assay, 20,000 cells were analyzed through a Cytoron Absolute cytometer (Ortho Diagnostic Systems).

3. Incubate the monolayer at 37°C for 30 min, in the dark.
4. Remove the dye-containing medium.
5. Incubate the monolayer with 5 ml of wash buffer (PBS or culture medium) at 37°C for 45 min, in the dark.
6. Remove the wash buffer.
7. Harvest the cells with 1 ml trypsin/EDTA (0.1% w/v:0.02% w/v) solution in PBS.
8. Resuspend in 1 ml PBS or culture medium containing a trypsin inhibitor (soybean trypsin inhibitor or bovine serum) and monodis-perse.

9. Add propidium iodide, at a final concentration of 1 μg/ml, before analysis.
10. Analyze the cell suspension using an argon ion laser tuned to 488 nm and measure the rhodamine 123 (green) fluorescence through a 530)5-nm band-pass filter.
11. Measure the red fluorescence through a 630-nm wavelength filter. Use a log scale to discriminate between dim signal owing to rhodamine 123 uptake by living cells and high signal owing to propidium uptake by dead cells (Fig. 4).

ACKNOWLEDGMENT

This work was supported by a grant from Région Rhône-Alpes and Université Joseph Fourier.

REFERENCES

1. C. A. Reinhardt, D. A. Pelli, and G. Zbinden, Interpretation of cell toxicity data for estimation of potential irritation, *Fundam. Chem. Toxicol. 23*:247 (1985).
2. A. P. Stammati, V. Silano, and F. Zucco, Toxicology investigations with cell cultures systems, *Toxicology 20*:19 (1981).
3. L. A. Kamentsky, M. R. Melamed, and H. Derman, Spectrophotometer: new instrument for ultrarapid cell analysis, *Science 150*:630 (1965).
4. P. K. Horan and M. R. Loken, A practical guide for the use of flow systems, *Flow Cytometry: Instrumentation and Data Analysis* (M. A. Vandilla, P. N. Dean, and O. D. Laerum, eds.), Academic Press, New York, 1985, p. 259.
5. M. R. Loken, R. G. Sweet, and L. A. Herzenberg, Cell discriminating by multiangle light scattering, *J. Histochem. Cytochem. 24*:284 (1976).
6. E. Combrier, P. Metezeau, X. Ronot, H. Gachelin, and M. Adolphe, Flow cytometric assessment of cell viability: a multifaceted analysis, *Cytotechnology 2*:27 (1989).
7. A. M. Pappenheim, Experimental studies upon lymphocytes. I. The reaction of lymphocytes under various experimental conditions, *J. Exp. Med. 25*:633 (1917).
8. P. R. Roper and B. Drewinko, Comparison of in vitro methods to determine drug induced cell lethality, *Cancer Res. 36*:2182 (1976).
9. B. K. Bhuyan, B. E. Loughman, T. J. Fraser, and K. J. Dai, Comparison of different methods of determining cell viability after exposure to cytotoxic compounds, *Exp. Cell Res. 97*:275 (1976).
10. A. W. Krause, W. W. Carley, and W. W. Webb, Fluorescent erythrosin B is preferable to trypan blue as a vital exclusion dye for mammalian cells in monolayer culture, *J. Histochem. Cytochem. 32*:1084 (1984).
11. J. B. LePecq and C. Paoletti, A fluorescent complex between ethidium bromide and nucleic acids. Physical-chemical characterization, *J. Mol. Biol. 27*:87 (1967).
12. C. Nicolini, F. Kendall, R. Baserga, C. Dessaive, and J. Fried, The G0–G1 transition of WI38 cells. I. Laser microfluorometry studies, *Exp. Cell Res. 106*:111 (1977).

13. A. N. Glazer, K. Peck, and E. A. Mathies, A stable double-stranded DNA–ethidium homodimer complex: application to picogram fluorescence detection of DNA in agarose gels, *Proc. Natl. Acad. Sci. USA 87:*3851 (1990).

14. C. Bucana, I. Saiki, and R. Nayar, Uptake and accumulation of the vital dye hydroethidine in neoplastic cells, *J. Cytochem. Histochem. 34:1*109 (1986).

15. X. Ronot, L. Benel, M. Adolphe, and J. C. Mounolou, Mitochondrial analysis in living cells: the use of rhodamine 123 and flow cytometry, *Biol. Cell 57:*1 (1986).

16. C. Dive, H. Cox, J. V. Watson, and P. Workman, Polar fluorescein derivatives as improved substrate probes for flow cytoenzymological assay of cellular esterases, *Mol. Cell Probes 2:*131 (1988).

17. R. P. Haugland, Fluorescent dyes for assessing vital cell functions, *Handbook of Fluorescent Probes and Research Chemicals* (K. D. Larison, ed.), Molecular Probes, Eugene, OR, 1992, p. 172.

18. L. Benel, X. Ronot, M. Kornprobst, M. Adolphe, and J. C. Mounolou, Mitochondrial uptake of rhodamine 123 by articular chondrocytes, *Cytometry 7:*281 (1986).

19. K. H. Jones and J. A. Senft, An improved method to determine cell viability by simultaneous staining with fluorescein diacetate–propidium iodide, *J. Histochem. Cytochem. 33:*77 (1985).

20. J. Tao, E. S. Crister, and J. K. Critser, Evaluation of mouse sperm acrosomal status and viability by flow cytometry, *Mol. Reprod. Dev. 36:*183 (1993)

21. K. A. Muirhead, E. D. Kloszewski, L. A. Antell, and D. E. Griswold, Identification of live cells for flow cytometric analysis of lymphiod subset proliferation in low viability populations, *J. Immunol. Methods 77:*77 (1985).

22. M. C. Riedy, K. A. Muirhead, C. P. Jensen, and C. C. Stewart, Use of a photolabeling technique to identify nonviable cells in fixed homologous or heterologous cell populations, *Cytometry 12:*133 (1991).

23. B. D. Jensen and K. A. Muirhead, Retrospective viability analysis using ethidium monoazide, *Methods in Nonradioactive Detection* (G. C. Howard, ed.), Appleton & Lange, Norwalk, CT, 1993, p. 329.

24. L. W. M. M. Terstappen, V. O. Shah, M. P. Conrad, D. Recktenwald, and M. R. Loken, Discriminating between damaged and intact cells in fixed flow cytometric samples, *Cytometry 9:*477 (1988).

25. J. K. A. Nicholson, S. W. Browning, S. L. Orloff, and J. S. McDougal, Inactivation of HIV-infected H9 cells in whole blood preparations by lysing/fixing reagents used in flow cytometry, *J. Immunol. Methods 160:*215 (1993).

26. M. E. Lalande, V. Ling, and R. G. Miller, Hoechst 33342 dye uptake as a probe of membrane permeability changes in mammalian cells, *Proc. Natl. Acad. Sci. USA 78:*363 (1981).

27. S. Paillasson, J. M. Millot, M. Manfait, and X. Ronot, DNA analysis in living cells: cytometric approaches, *Visualization of Nucleic Acids* (G. Morel, ed.), CRC Press, New York, 1995 (in press).

28. C. Waymouth, To disaggregate or not to disaggregate; injury and cell disaggregation, transient or permanent? *In Vitro 10:*97 (1974).

11

Analysis of Apoptosis
by Flow Cytometry

ANNE E. MILNER, HONG WANG, AND CHRISTOPHER D. GREGORY
The University of Birmingham,
Birmingham, England

I. INTRODUCTION

A. Apoptosis: Physiological Cell Death

The term *apoptosis* describes a series of gross cellular changes characteristic of an active or programmed form of cell death that is now accepted as a physiological, as opposed to accidental, mode of cell deletion. Active cell death represents a normal control mechanism as profoundly important to multicellular organisms as cell proliferation or differentiation. In mammals, the significance of physiological cell death is illustrated by the removal of cells through activation of their apoptotic mechanism during tissue modeling and organogenesis in embryonic development, and in the control of homeostasis in a diversity of tissue types [1]. There is much evidence to support the view that programmed cell death is a cell-autonomous process [2]. It has also been suggested that, with few exceptions, all cells from all lineages are programmed to undergo active self-destruction unless they receive survival signals to suppress that program [3].

Apoptosis is a genetically regulated, thermodynamically demanding process that contrasts markedly with the passive mode of cell death, necrosis.

Necrotic cells play no active role in their own demise, and the loss in membrane integrity that occurs during necrosis leads to leakage of intracellular contents and subsequent histotoxic damage and inflammation. In vivo, the process of apoptosis is such that membrane integrity is retained until the dying cells have been cleared by phagocytosis. In vitro death by apoptosis in populations lacking phagocytes ultimately results in plasma membrane breakdown, a phenomenon sometimes referred to as *secondary necrosis*. Although apoptosis has yet to be ascribed a definitive biochemical mechanism, activation of the apoptotic machinery induces a distinct series of gross structural changes, identifiable by microscopy, including condensation, margination, and fragmentation of chromatin; cell shrinkage associated with cytoplasmic condensation; and dilation of endoplasmic reticulum. These changes may be accompanied by the fragmentation of the cell into membrane-bound apoptotic bodies containing cytoplasmic organelles or nuclear components, or both [4].

B. Classic Methods of Detecting Apoptosis

1. Microscopy

Apoptosis is most reliably identified by microscopy. Transmission electron microscopy has provided a wealth of morphological information on the consequences of apoptosis (Fig. 1) [5]. Light microscopy of tissue sections or cell smears stained by standard methods is widely used for qualitative and quantita-

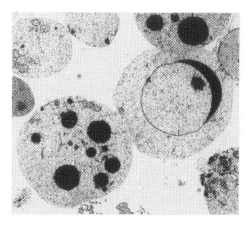

FIGURE 1 Electron microscopy of Burkitt lymphoma cells induced into apoptosis by incubation for 5 h on ice, followed by 5 h at 37°C. Condensed and fragmented chromatin can be identified.

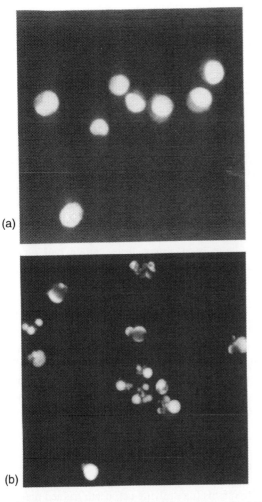

FIGURE 2 Fluorescence light microscopy of viable (a) and apoptotic (b) Burkitt lymphoma cells. Apoptosis was induced as in Figure 1. Cells were stained with acridine orange.

tive analyses of apoptosis [6,7], as has fluorescence light microscopy of cells stained with DNA-binding fluorochromes, which allows the visualization of condensed and fragmented chromatin, cellular shrinkage, and apoptotic bodies (Fig. 2) [8,9]. End-labeling of damaged DNA (see next section) has also been used to tag apoptotic cells for light microscopic analysis. However, in the absence of complex image analysis systems and superhuman efforts, the obvious limitation of microscopy-based techniques in quantitative studies of apoptosis lies in the relatively small fraction of cells in a given population that can be evaluated effectively.

2. Internucleosomal DNA Fragmentation

Seminal studies by Wyllie and colleagues showed that apoptosis in rodent thymocytes was accompanied by endonucleolytic DNA cleavage [10], and DNA fragmentation has since been widely used as an assay of apoptosis in isolated cell populations. Typically, the morphological changes characteristic of apoptosis are associated with the generation of multiple oligonucleosomal fragments that appear after electrophoresis as a "ladder" pattern, separated by spaces equivalent to the internucleosomal distance of about 200 bp. However, such patterns have also been observed in DNA isolated from necrotic cells [11]. Conversely, oligonucleosomal ladders have been absent from certain cell types undergoing classic, morphologically defined apoptosis, although all types of apoptotic cells appear to generate DNA fragments of a higher order (300–50 kbp)[12].

Determination of the ratio of fragmented versus total DNA in whole-cell lysates by high-speed centrifugation has been performed in numerous studies [13,14] as a method of quantifying apoptosis in cell populations in vitro. The clear drawbacks of using DNA fragmentation methods in assessing apoptosis are (a) that relatively large cell numbers are required, (b) that fragmentation patterns in gels are largely nonquantitative or, at best, semi-quantitative, and (c) that the methods are limited in providing information about only whole populations, rather than individual cells.

C. Advantages of Flow Cytometry

The widespread use of in vitro cell systems for studying cell death has led many researchers to turn to flow cytometry as an alternative to the classic methods of analyzing apoptosis. Flow cytometry offers many advantages over the methods summarized in the foregoing, including the capability to analyze individual cells, ease of sample preparation, rapidity of sample analysis, and low cell number requirement. In addition it allows cell subpopulations to be identified by phenotypic characteristics (an important consideration in heterogeneous populations) and permits the simultaneous measurement of several cellular parameters. Here, we will review several flow cytometric techniques, of varying complexity, that have been developed to detect cellular and subcellular changes occurring during apoptosis.

II. CELLULAR VOLUME AND CELLULAR DENSITY CHANGES BY TWO-DIMENSIONAL LIGHT SCATTER ANALYSIS

A. Light Scatter Characteristics of Apoptotic Cells

In unstained cell populations, changes in cellular morphology can be detected by flow cytometry on the basis of light scatter measurements. Two parameters can be assessed: forward-angle light scatter and light scattered at 90°. As light scattered by a particle in the forward direction (low angle) is proportional to the size of that article, and light scattered at larger angles is sensitive to internal structure of the particle [15], analysis of forward and 90° light scatter has been used to reveal information on cellular size and density. Several laboratories, including ours, have studied to what extent the reduction in cellular volume and increases in cellular density that are observed during apoptosis are revealed by alterations in the light-scatter properties of the dying cells.

Studies of lymphocytes (freshly isolated, primary cultures or cell lines) have indicated that induction of apoptosis (as assessed by classic criteria) results in the accumulation of a population with reduced forward light scatter (FLS) and enhanced 90° light-scatter characteristics when analyzed on a two-dimensional dot plot (Fig. 3a) [16–18]. Cell sorting on the basis of light-scatter properties and

(a) **(b)**

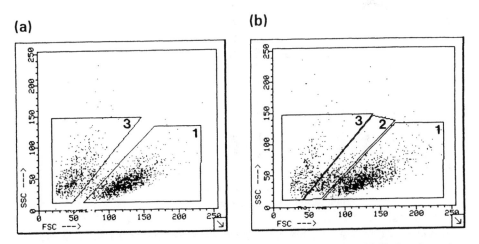

FIGURE 3 Forward light scatter (FSC) versus side light scatter (SSC) dot plots of Burkitt lymphoma cells undergoing apoptosis induced by incubation with the calcium ionophore, ionomycin. Region 1, viable cells; region 2, early apoptotic cells; region 3, late apoptotic cells.

subsequent staining and DNA electrophoresis of the sorted cells reveals formally that this population is composed of apoptotic cells. Thus, flow cytometry can reveal apoptotic cells in unstained preparations solely on the basis of the decrease in FLS combined with the increase in 90° light scatter that occurs as a result of cellular shrinkage and increased internal granularity or density. This assay has been effective with either viable or fixed cells, making it convenient for routine analyses. Additionally, with some cell types, it may be possible to detect different stages of apoptosis by different positions on the two-dimensional (2-D) light-scatter plots. Most apoptotic cells are clearly smaller and more dense than the viable cells, but some, possibly representing cells at an earlier stage in the apoptotic process, show only slight differences in light scatter, compared with the viable population (see Fig. 3b).

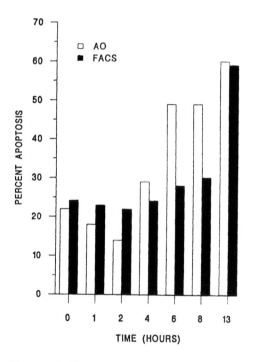

FIGURE 4 Comparison of the percentage of apoptosis detected in Burkitt lymphoma cells measured either by fluorescence microscopic analysis of acridine orange-stained cells (AO) or by flow cytometric analysis of forward light scatter versus side light scatter dot plots (FACS). Apoptosis was induced by incubation with ionomycin for varying time periods.

B. Assay Limitations

On a cautionary note, the changes in light-scatter properties that accompany apoptosis lag behind the morphological changes detected by acridine orange staining. Thus, comparison of data obtained from light-scatter and aridine orange measurements (Fig. 4) of a population in which apoptosis is on-going indicate that there is a short delay in the shift of apoptotic cells from the viable to the dead zone in the flow cytometric assay: at early time points the number of apoptotic cells in the dead zone is an underestimate of the true size of the population. At the endpoint, however, the light-scatter assay provides comparable (and far more speedily obtained) data to that obtained after manual counting of acridine orange-stained cells.

Although cells undergoing primary necrosis generally display lower 90° light-scatter properties than apoptotic cells, necrotic populations may impinge on the apoptotic zone. Therefore, caution must be exercised when assessing the behavior of a new cell type or a new apoptosis-inducing treatment. Furthermore, to ensure that the changes in light scatter that occur in apoptotic cells are not masked by the light-scatter profiles (viable or otherwise) from a second cell type, the method also tends to be limited to populations that, in their viable state, show little size variation and consist predominantly of a single cell type. As a general rule, it can be concluded that, although the light-scatter method is an excellent tool for quantifying apoptosis in isolated cell samples, it requires a relatively pure, homogeneous starting population and should be used only after diagnosis of apoptosis by a complementary technique, such as acridine orange staining.

III. ASSESSMENT OF DNA CONDENSATION AND FRAGMENTATION BY SINGLE-PARAMETER FLUORESCENCE ANALYSIS

A. The Subdiploid DNA Fluorescence Peak

Several studies have shown that changes in chromatin conformation can be reflected by alterations in accessibility of fluorochromes to DNA [19]. Thus, during cell differentiation and, in some circumstances, chromatin condensation during the cell cycle, altered accessibility to DNA of several dyes can be measured as changes in fluorescence intensity. This has led many workers to assess to what extent the condensation and fragmentation of DNA that occurs during apoptosis can be detected as changes in the fluorescence emitted by DNA-binding fluorochromes. A number of fluorochromes that either intercalate in the groove between the two strands of the DNA double-helix (e.g., propidium iodide, ethidium bromide, acridine orange, dactinomycin, (actinomycin D)], or bind externally to DNA [e.g., Hoechst 33342, plicamycin, (mithramycin), chromomycin A_3)] have been tested in this type of analysis.

Single-parameter analysis of the fluorescence emitted from DNA-binding fluorochromes in a normal-cycling population allows the cell cycle stages of individual cells to be revealed. The G_0/G_1 phase is represented by a narrow peak of fluorescence from cells with diploid DNA. The S-phase cells, which are synthesizing new DNA, form a diffuse plateau of increasing fluorescence, and cells in G_2/M, which have completed DNA duplication, form a narrow peak in which the mean fluorescence intensity is double that of the G_1 peak (Fig. 5a). Analysis of thymocytes undergoing apoptosis demonstrated that, with both intercalating and externally binding dyes, a subdiploid DNA peak (such as that shown in Fig. 5b) was present in addition to the normal diploid G_0/G_1 peak [20]. These dyes included acridines, actinomycins, chromomycinones, anthracycline, Hoechst dyes, and 4', 6-diamidino-2-phenylindole (DAPI), and there was close accordance in the percentages of cells in the subdiploid peak for all the dyes, indicating that apoptotic cells were always defined by a discrete decrease in fluorescence intensity.

The subdiploid DNA peak has been observed in many different cell types undergoing apoptosis. Thus, murine interleukin-3 (IL-3)-dependent BAF-3 cells, deprived of IL-3; mouse thymocytes incubated with glucocorticoids; MCF-7 breast cancer cells, treated with the somatostatin analogue, octreotide (SMS 201-995); peripheral blood lymphocytes, treated with various drugs; mouse splenic B lymphocytes, maintained in culture; and mouse bone marrow cells, treated with

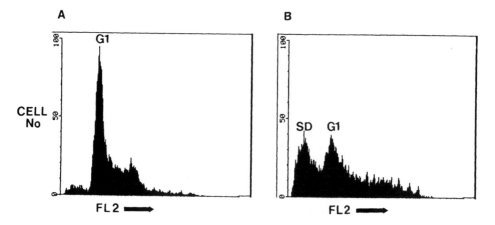

FIGURE 5 Cell cycle analysis of (A) viable and (B) apoptotic Burkitt lymphoma cells. The G_1 and subdiploid (SD) peaks can be identified. Apoptosis was measured 24 h after irradiation with 4 gray.

dexamethasone, all display a subdiploid DNA peak, after treatment, when stained with propidium iodide or acridine orange [18, 21–29]. Sorting of cells within the subdiploid peak by FACS has confirmed that they are indeed apoptotic. Likewise, the subdiploid DNA peak is readily demonstrable in apoptotic cells purified on Percoll gradients [23].

The observed reduction in fluorescence emission of DNA-binding dyes in apoptotic cells implies that such cells either contain reduced levels of DNA (presumably as a result of endonuclease activity and subsequent leakage of small DNA fragments), or contain fluorochrome-inaccessible DNA (caused by chromatin condensation). Available evidence supports each of these alternatives. Thus, a subdiploid population of thymocytes emerged after 4 h of treatment with dexamethasone, even though no DNA had been lost from the nuclei, indicating that a portion of the chromatin in apoptotic cells had condensed to a state of reduced accessibility [26]. Several observations indicate that reduced DNA stainability is a consequence of partial loss of DNA from apoptotic cells owing to activation of an endogenous endonuclease and subsequent diffusion of the low molecular weight fragments from the cell before cell measurement [19]. Notably, the decrease in DNA stainability has been reported to be dependent on the storage time of sample before analysis. Thus, analysis of apoptotic cells, fixed in cold ethanol revealed that the subdiplod peak became more pronounced with increasing storage time in phosphate-buffered saline (PBS). Furthermore, the presence of low molecular weight DNA of the size equivalent to mononucleosomes was detected in the PBS in which the apoptotic cells were suspended [19]. In the same study, apoptotic cells stained with DAPI initially contained DNA evenly dispersed in the cytoplasm when viewed by UV light microscopy. However, when apoptotic cells fixed on slides were subsequently rinsed for more than 10 min in PBS before staining, the DNA in the cytoplasm was no longer apparent [19].

B. Assay Limitations

Although monoparametric analysis of DNA-binding fluorochromes can clearly define apoptotic cells containing subdiploid levels of DNA, the assay has several limitations that should be considered when interpreting data. First, in a cycling population, cells undergoing apoptosis from S or G_2/M phases of the cell cycle also emit lower levels of DNA fluorescence than their viable counterparts. Such cells merge with the viable G_1- and S-phase populations and, consequently, remain undetected [23]. This underestimation of the true size of the apoptotic population can be seen readily when the percentage of cells in the subdiploid peak is compared with the percentage of apoptotic cells scored by microscopic counting after acridine orange staining (Table 1). Therefore, this technique is far more reliable when applied either to noncycling populations or to populations in

TABLE 1   Quantitation of Apoptosis in Burkitt Lymphoma Cells by Fluorescence Microscopy After Acridine Orange Staining and Flow Cytometric Evaluation of Subdiploid DNA

	Percent apoptosis	
Apoptotic trigger	Acridine orange	Subdiploid DNA
Monoclonal anti-IgM antibody	60	27
Ionomycin	78	37

which entry into apoptosis is restricted to the G_1 phase of the cell cycle. A second note of caution relates to the possibility that decreased stainability with DNA fluorochromes also occurs during cell degeneration, such as necrosis, despite reports that necrosis, induced in mouse thymocytes or MCF-7 cells by 0.1% sodium azide, does not result in the generation of a subdiploid DNA peak [24,25]. Nonspecific degradation of DNA in necrotic cells is a potentially significant source of contamination of the subdiploid apoptotic population, particularly in samples containing mixtures of cells undergoing primary necrosis in addition to apoptosis. Again, as concluded in Section II, additional (most reliably morphological) methods of diagnosing apoptosis are an essential accompaniment to this technique, at least in preliminary investigations of new cell systems or treatments.

IV. MULTIPARAMETER LIGHT SCATTER AND DNA FLUORESCENCE ANALYSIS

A. Membrane-Impermeable DNA-Binding Dyes

Many of the limitations of monoparametric fluorescence analysis described in the foregoing may be overcome by combining measurements of DNA fluorescence with those of cell size, as assessed by forward light scatter (FLS). For example, dual-parameter analysis of DNA condensation or fragmentation and FLS of cells stained with propidium iodide (after permeabilization) permits an accurate assessment of cells entering apoptosis from multiple stages of the cell cycle. As shown in Figure 6, such cells display a decrease in DNA fluorescence intensity, combined with a decrease in cell size and, consequently, are represented as a distinct population that can be readily quantified.

Simultaneous analysis of FLS and fluorescence emission from membrane-impermeable DNA-binding dyes, such as propidium iodide or ethidium bromide, has been used to monitor the kinetics of the progression of apoptosis. For example, distinct populations of early and late apoptotic thymocytes were observed to

A B

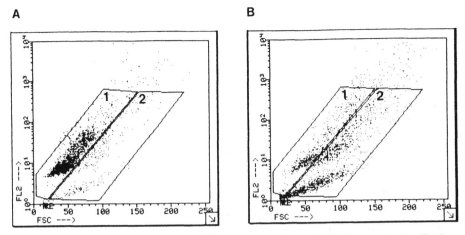

FIGURE 6 Forward light scatter (FSC) versus propidium iodide fluorescence (FL2) dot plots of (A) viable and (B) apoptotic Burkitt lymphoma cells. Cells induced into apoptosis by irradiation with 4 gray demonstrate decreased binding of propidium iodide simultaneously with a decrease in cell size. Region 1, viable cells; region 2, apoptotic cells.

accumulate after methylprednisolone treatment [30]: the earliest population was characterised by low FLS and low ethidium staining, a later population displayed an increase in ethidium fluorescence as the cell membranes became permeable to the dye, and the latest cells progressed to lose the ethidium-bound DNA.

B. Hoechst 33342

An alternative DNA-binding fluorochrome that has been used in combination with FLS measurements in the visualization of apoptotic cells is Hoechst 33342. This is an externally binding DNA dye that readily diffuses into intact viable cells and apoptotic cells. Dual-parameter analysis of fluorescence intensity and FLS of Hoechst-stained thymocytes or Burkitt lymphoma cells has shown that, after induction of apoptosis, a distinct population accumulates that displays low FLS and high fluorescence properties [17,31–32]. Confirmation that this population is composed of apoptotic cells was obtained by cell-sorting. This method, therefore, permits the visualization of apoptotic cells based on the combination of increased Hoechst fluorescence of the DNA and decreased FLS of the cell body. When using this assay, care should be exercised in controlling both the dye concentration and the labeling time, as extended labeling may result in a higher fluorescence intensity of viable cells, resulting in a diminished difference between viable and apoptotic populations [32].

The mechanism by which Hoechst preferentially stain cells undergoing DNA fragmentation is unknown. It is known that this benzimidazole dye binds to external DNA domains in minor grooves. It is possible, therefore, that DNA fragmentation exposes more of the A-T base pairs (to which benzimidazole dyes preferentially bind), resulting in a higher concentration of stain in apoptotic cells. Alternatively, or additionally, the increase in DNA fluorescence may reflect a decrease in the ability of apoptotic cells to actively efflux the probe, a precedent that has been noted in multidrug-resistant cells. Thus, it is conceivable that, as cells undergo apoptosis, the efflux mechanism may be inhibited or lost. Preliminary data suggest that higher Hoechst fluorescence intensity may be due to an increase in cell membrane permeability [32]; therefore, care may be needed in assessing apoptosis-triggering agents that alter membrane permeability. In addition, deliberate cell fixation and permeabilization may change the route of dye entry, allowing cells that were viable at the time of fixation or permeabilization to bind high concentrations of the dye. In this scenario, apoptotic cells may then appear as a subdiploid DNA-containing population as described in Section III.A.

Controlled staining with Hoechst 33342, which can enter the nucleus of nonpermeabilized cells, in combination with propidium iodide, which can access only the DNA of permeabilized cells, allows those cells that have undergone necrosis (either primary or secondary), with consequent membrane damage, to be discriminated from viable and apoptotic cells the membranes of which remain intact [17,31,32]. The appearance of secondary necrotic cells thus permits the late kinetics of the apoptotic process to be monitored in homogeneous cell populations. When Hoechst 33342 was combined with propidium iodide staining of unfixed BAF-3 cells (a murine IL-3–dependent B-cell line) cultured in the absence of IL-3, it was possible to identify cells in transit from the apoptotic to the dead (secondary necrotic) compartment, since, in the dead zone, the cells were permeable to propidium [27]. Within the dead compartment, decreased propidium iodide fluorescence identified cells with extensively degraded DNA. This method, therefore, permits differential quantification of apoptotic cells with either intact or leaky membranes.

C. Dual Fluorescence Analysis of DNA and RNA

The metachromatic nucleic acid-binding fluorochrome, acridine orange, has been used in the simultaneous analysis of DNA and RNA in apoptotic cells. This dye fluoresces green when bound to double-stranded nucleic acids and red when associated with single-stranded nucleic acids. In practice, acridine orange-stained cells emit green fluorescence from their nuclei and red from their cytoplasm when they are excited with blue light. Induction of apoptosis in a murine B-cell line generated a distinct population that, as assessed by acridine orange staining, contained little or no RNA and subdiploid DNA [33]. Since reduction in RNA can be used to assess cells undergoing apoptosis from late, as well as early, stages of the cell cycle, this method provides an opportunity to quantify apoptosis within cycling populations with greater accuracy than by measuring subdiploid DNA alone. However, as some denaturation of DNA may occur in apoptotic cells, the resulting single-stranded DNA can obscure the possible loss of RNA, thereby reducing the accuracy of the technique.

D. Dual Fluorescence Analysis of DNA and Protein

Differential staining of DNA and protein with DAPI and sulforhodamine 101, respectively, has revealed that certain apoptotic cells, such as those arising in the HL60 leukemia line, have markedly diminished levels of total protein. This change appears to occur in parallel with the decrease in DNA content [19]. Thus, the simultaneous measurement of DNA and protein can discriminate between live and apoptotic cells. However, a clear limitation of the assay is that primary necrotic cells also display reduced levels of protein, as well as of DNA.

V. FLUORESCENCE ANALYSIS OF DNA STRAND BREAKS

A. Application to Flow Cytometry

The DNA strand breaks that accompany apoptosis can be visualized by an in situ nick translation assay in which fluorochrome-labeled nucleotides are annealed onto the exposed DNA ends [34]. This method is highly amenable to flow cytometric analysis, and a variety of cell types stained with fluorescein isothiocyanate (FITC)-conjugated UTP displayed increases in FITC-generated fluorescence after induction of apoptosis, in a manner that correlated with the onset of apoptosis assessed microscopically and by DNA ladder formation (Fig. 7) [35–37]. Simultaneous measurement of DNA content by propidium iodide staining that gives rise to a subdiploid peak has permitted correlation of the appearance of DNA breaks with the position of the cell in the cycle and has demonstrated that DNA breaks are detectable before the loss of DNA staining by propidium iodide [35]. This technique can be combined readily with immunocytochemistry of surface or cytoplasmic antigens for analysis of apoptosis in subsets of cells in mixed populations.

B. Assay Limitations

Although the incorporation of UTP provides evidence of DNA strand breaks, it is not proof of apoptosis. For example, alkylating agents that are used in treatment regimens for CLL and lymphoma can induce specific DNA damage in cells

A   B

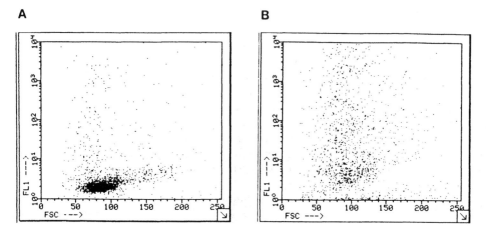

FIGURE 7 Forward light scatter (FCS) versus FITC fluorescence (FL1) of (A) viable and (B) apoptotic Burkitt lymphoma cells. Cells induced into apoptosis by irradiation with 4 gray demonstrate an increased binding of FITC-conjugated UTP.

that escape apoptotic death [28]. It should also be noted that in situ end-labeling may also tag necrotic cells with high efficiency. In addition, it is a theoretical possibility that end-labeling may also detect nonlinked DNA fragments occurring during DNA replication, repair, transcription, or recombination. In practice, proliferating cells remain negative after staining, probably because the signal obtained by labeling of single, nonlinked DNA fragments is too weak to exceed the background autofluorescence threshold [35,36], Further evidence that single-strand breaks label too weakly to exceed background fluorescence levels is provided by observations of Burkitt lymphoma cells that, when labeled immediately after low-dose γ-irradiation, fail to fluoresce above background. [A. E. Milner and C. D. Gregory, unpublished data].

VI. SUMMARY AND CONCLUSIONS

Specific changes in cell structure that accompany the process of apoptosis in a diverse spectrum of cell types can be detected by flow cytometry. Loss of cell volume and increased cell density can be revealed by simple light scatter analysis. DNA condensation and fragmentation can be detected as alterations in DNA-binding fluorochromes. Multiparameter light scatter and fluorescence analyses can be carried out simultaneously to quantify apoptotic cells with greater accuracy and to dissect out discrete stages in the apoptotic process.

In conclusion, flow cytometry provides a powerful tool to quantify apoptosis in isolated cell populations. However, no flow cytometric-based assay is without significant limitation: for example, two-parameter light-scatter analysis of unstained cells is less accurate than microscopic analysis of acridine orange-stained cells for assessment of the early stages in the apoptotic process; single-parameter fluorescence analysis of DNA changes can be reliably applied to only noncycling populations; analysis of strand breaks is not proof of apoptosis, since strand breaks can occur in many situations, indeed in the absence of cell death; necrosis, be it primary or secondary, is a source of inaccuracy for all methods. Even though the choice of technique may be restricted by the capacity of the hardware available for analysis, increased sophistication does not guarantee increased accuracy. No flow cytometric technique, however sophisticated, has yet mastered the art of formal diagnosis of apoptosis. Therefore, for uncharacterized systems, changes observed on the basis of flow cytometric data should be carefully monitored, at least initially, by light microscopy.

ACKNOWLEDGMENTS

The work in this laboratory is supported by the Cancer Research Campaign and the Leukaemia Research Foundation

REFERENCES

1. A. H. Wyllie, Apoptosis and the regulation of cell numbers in normal and neoplastic tissues: an overview, *Cancer Metastasis Rev. 11*:95 (1992).
2. Z. N Oltvai and S. J. Korsemeyer, Checkpoints of dueling dimers foil death wishes, *Cell 79*:189 (1994).
3. M. C. Raff, B. A. Barres, J. F. Burne, H. S. Coles, Y. Ishizaki, and M. D. Jacobson, Programmed cell death and the control of cell survival: lessons from the nervous system, *Science 262*:695 (1993).
4. A. H. Wyllie, Apoptosis: cell death under homeostatic control, *Mech. Mod. Toxicol. Arch. Toxicol. Suppl. 11*:3 (1987).
5. A. H. Wyllie, G. J. Beattie, and A. D. Hargreaves, Chromatin changes in apoptosis, *Histochem. J. 13*:681 (1981).
6. Y. J. Liu, D. E. Joshua, G. T. Williams, C. A. Smith, J. Gordon, and I. C. M. MacLennan, Mechanism of antigen-driven selection in germinal centres, *Nature 342*:6252 (1989).
7. A. J. Merritt, C. S. Potten, C. J. Kemp, J. A. Hickman, A. Balmain, D. P. Lane, and P. A. Hall, The role of p53 in spontaneous and radiation-induced apoptosis in the gastrointestinal tract of normal and p53 deficient mice, *Cancer Res. 54*:614 (1994).
8. C. D. Gregory, C. Dive, S. Henderson, C. A. Smith, G. T. Williams, J. Gordon, and A. B. Rickinson, Activation of Epstein-Barr virus latent genes protects human B cells from death by apoptosis, *Nature 349*:612 (1991).

9. S. L. Silins and T. B. Sculley, Burkitt's lymphoma cells are resistant to programmed cell death in the presence of the EBV latent antigen EBNA-4, *Int. J. Cancer 60*:65 (1995).

10. M. J. Arends, R. G. Morris, and A. H. Wyllie, Apoptosis: the role of the endonuclease, *Am. J. Pathol. 136*:593 (1990).

11. R. J. Collins, B. V. Harmon, G. C. Gobe, and J. F. Kerr, Internucleosomal DNA cleavage should not be the sole criterion for identifying apoptosis, *Int. J. Radiat. Biol. 61*:451 (1992).

12. F. Oberhammer, J. W. Wilson, C. Dive, I. D. Morris, J. A. Hickman, A. E. Wakeling, P.R. Walker, and M. Sikorska, Apoptotic death in epithelial cells: cleavage of DNA to 300 and/or 50 kb fragments prior to or in the absence of internucleosomal fragmentation, *EMBO J. 12*:3679 (1993).

13. R. C. Duke, R. Chervenak, and J. J. Cohen, Endogenous endonuclease-induced DNA fragmentation: an early event in cell-mediated cytolysis, *Proc. Natl. Acad. Sci. USA 80*:6361 (1983).

14. R. C. Duke and J. J. Cohen, IL-2 addiction: withdrawal of growth factor activates a suicide program in dependent T cells, *Lymphokine Res. 5*:289 (1986).

15. G. C. Salzman, Light scattering and flow cytometry, *Flow Cytometry and Sorting*, 9th ed. (M. R. Melamed, ed.), Wiley-Liss, New York, 1990, p. 81.

16. W. Swat, L. Ignatowicz, and P. Kisielow, Detection of apoptosis of immature CD4+8+ thymocytes by flow cytometry, *J. Immunol. Methods 137*:79 (1991).

17. C. Dive, C. D. Gregory, D. J. Phipps, D. L. Evans, A. E. Milner, and A. H. Wyllie, Analysis and discrimination of necrosis and apoptosis by multiparameter flow cytometry, *Biochim. Biophys. Acta 1133*:275 (1992).

18. V. A. Illera, C. E. Perandones, L. L. Stunz, D. A. Mower, and R. E. Ashman, Apoptosis in splenic B lymphocytes, *J. Immunol. 151*:2965 (1993).

19. Z. Darzynkiewicz, S. Bruno, G. Del Bino, W. Gorczyca, M. A. Hotz, P. Lassota, and F. Traganos, Features of apoptotic cells measured by flow cytometry, *Cytometry 13*:795 (1992).

20. W. G. Telford, L. E. King, and P. J. Fraker, Comparative evaluation of several DNA binding dyes in the detection of apoptosis-associated chromatin degradation by flow cytometry, *Cytometry 13*:137 (1992).

21. G. Rodriguez-Tarduchy, M. Collins, and A. Lopez-Rivas, Regulation of apoptosis in interleukin-3-dependent hemopoietic cells by interleukin-3 and calcium ionophores, *EMBO J. 9*:2997 (1990).

22. F. Ojeda, M. I. Guarda, C. Maldonado, and H. Folch, Protein kinase-C involvement in thymocyte apoptosis induced by hydrocortisone, *Cell. Immunol. 125*:535 (1990).

23. W. G. Telford, L. E. King, and P. J. Fraker, Evaluation of glucocorticoid-induced DNA fragmentation in mouse thymocytes by flow cytometry, *Cell Prolif. 24*:447 (1991).

24. I. Nicoletti, G. Migliorati, M. C. Pagliacci, F. Grignani, and C. Riccardi, A rapid and simple method for measuring thymocyte apoptosis by propidium iodide staining and flow cytometry, *J. Immunol. Methods 139*:271 (1991).

25. M. C. Pagliacci, R. Tognellini, F. Grignani, and I. Nicoletti, Inhibition of human breast cancer cell (MCF-7) growth in vitro by the somatostatin analog SMS 201-

995: effects on cell cycle parameters and apoptotic cell death, *Endocrinology* *129*:2555 (1991).

26. P. R. Walker, C. Smith, T. Youdale, J. Leblanc, J. F. Whitfield, and M. Sikorska, Topoisomerase II-reactive chemotherapeutic drugs induce apoptosis in thymocytes, *Cancer Res. 51*:1078 (1991).

27. M. G. Ormerod, M. K. L. Collins, G. Rodriguez-Tarduchy, and D. Robertson, Apoptosis in interleukin-3-dependent haemopoietic cells, *J. Immunol. Methods 153*:57 (1992).

28. O. S. Frankfurt, J. J. Byrnes, D. Seckinger, and E. V. Sugarbaker, Apoptosis (programmed cell death) and the evaluation of chemosensitivity in chronic lympho-cytic leukaemia and lymphoma, *Oncol. Res. 5*:37 (1993).

29. B. A. Garvy, W. G. Telford, L. E. King, and P. J. Fraker, Glucocorticoids and irradi-ation-induced apoptosis in normal murine bone marrow B-lineage lymphocytes as determined by flow cytometry, *Immunology 76*:270 (1993).

30. C. L. P. Deckers, A. B. Lyons, K. Samuel, A. Sanderson, and A. H. Maddy, Alter-native pathways of apoptosis induced by methylprednisolone and valinomycin anal-ysed by flow cytometry, *Exp. Cell Res. 208*:362 (1993).

31. J. A. Hardin, D. H. Sherr, M. A. DeMaria, and P. A. Lopez, A simple fluorescence method for surface antigen phenotyping of lymphocytes undergoing DNA fragmen-tation, *J. Immunol. Methods 154*:99 (1992).

32. G. M. Cohen, X.-M. Sun, R. T. Snowden, D. Dinsdale, and D. N. Skilleter, Key morphological features of apoptosis may occur in the absence of internucleosomal DNA fragmentation, *Biochem. J. 286*:331 (1992).

33. T. L. Jones and D. Lafrenz, Quantitative determination of the induction of apoptosis in a murine B cell line using flow cytometry bivariate cell cycle analysis, *Cell. Immunol. 142*:348 (1992).

34. R. R. Jonker, J. G. J. Bauman, and J. M. W. Visser, Detection of apoptosis using nonradioactive in situ nick translation, *NATO Advanced Study Institutes Pro-gramme. New Developments in Flow Cytometry. Lectures Option C*, CNRS, Ville-juif, 1992, p. 30.

35. W. Gorczyca, M. R. Melamed, and Z. Darzynkiewicz, Apoptosis of S-phase HL-60 cells induced by topoisomerase inhibitors: detection of DNA strand breaks by flow cytometry using the in situ nick translation assay, *Toxicol. Lett. 67*:249 (1993).

36. R. Gold, M. Schmied, G. Rothe, H. Zischler, H. Breitschhopf, H. Wekerle, and H. Lassmann, Detection of DNA fragmentation in apoptosis: application of in situ nick translation to cell culture systems and tissue sections, *J. Histochem Cytochem. 41*:1023 (1993).

12

Monitoring Baculovirus Infection and Protein Expression in Insect Cells

Athanassia K. Kioukia
"Demokritos" National Centre for Scientific Research,
Athens, Greece

N. H. Simpson, Jatin D. Shah, A. Nicholas Emery, Mohamed Al-Rubeai
The University of Birmingham,
Birmingham, England

I. INTRODUCTION

The insect cell–baculovirus system has been widely used either for insecticidal purposes (when cells are infected with wild-type virus) or for foreign gene expression (when cells are infected with a recombinant virus). The latter has broad applicability as an alternative to prokaryotic or other eukaryotic expression systems and has shown great promise recently for the production of biologically active proteins [1]. The operation of this system for large-scale production, however, is rather complex and requires progression through three different stages: the growth of cells, infection with virus, and protein expression. As process development has moved toward the increase of culture productivity, there has been an increased need for quick and accurate information on the state of the culture at every stage. Given such information, it should then be the aim at least to predict the performance of subsequent infection and protein expression and, ideally, to manipulate physicochemical parameters in favor of improved production.

Conventional methods of monitoring cell growth and infection with baculovirus in insect cells involve microscopic counting, for example, of cells and viral inclusion bodies (polyhedra), when present, or biochemical assaying (i.e., metabolic assays, virus titer plaque assay, or other). Viability assessment by trypan blue or other exclusion stains can, however, be misleading as a means of characterizing the cells metabolic state [2]. Polyhedra counting of infected cells can be inadequate, as it is restricted by a two-dimensional optical spectrum as well as being applicable to only wild-type virus. Following infection with recombinant baculovirus expressing heterologous protein, protein expression is assessed by tedious and labor-intensive biochemical means that monitor only average population yields. The use of flow cytometry (FC), on the other hand, has the potential to reveal many otherwise unknown properties by examining the intrinsic properties of individual cells. For instance, cells undergoing any kind of stress are expected to respond through their DNA and RNA or mitochondrial activity, and these are parameters measurable by FC. Some FC applications for the identification of insect cell infection with baculovirus and protein expression exist [3–6], and extension of the application of FC to process identification and control was first demonstrated by Miltenburger's group at the Technical University of Darmstadt [reported in Ref. 6]. They used FC measurement of the proportion of cells in the G_1 phase to drive the nutrient feed rate strategy in a perfusion culture of insect cells. Here we report further exploitation of the potential of FC to closely monitor viral infection and protein expression in insect cell-baculovirus cultures.

II. MATERIALS AND METHODS

Spodoptera frugiperda (Sf9) insect cells were cultivated in small spinner flasks (50–100 ml) in TC-100 medium, with 5% fetal calf serum (FCS) at 28°C. Cells were magnetically stirred at a rate of 120 rpm with small magnetic followers. Following growth, cells were infected in their flasks with wild-type *Autographa californica* nuclear polyhedrosis virus (AcNPV) or recombinant type virus expressing the enzyme β–galactosidase [7]. The Sf9 cells and the wild-type baculovirus were obtained from the Cancer Studies Department in the University of Birmingham, United Kingdom, and the recombinant virus strain came from Professor H. G. Miltenburger, Technical University, Darmstadt, Germany.

From the these cultures regular samples of cells were either fixed in cold 70% ethanol for at least 30 min at 4°C or used as live cultures. Parameters studied were DNA contents, light-scattering properties, and intracellular β–galactosidase contents. All samples were adjusted to a standard cell concentration of 5 × 10⁵/ml before staining and analysis.

A. Flow Cytometric Analysis

Fixed cells in 70% ethanol were initially washed twice in phosphate-buffered saline (PBS) and then incubated with RNase for 30 min at 37°C to remove all dsRNA. Excess RNase was washed out after centrifugation at 1000 rpm for 5 min and resuspension of the cell pellet twice in PBS. Before the analysis, cells were stained with 50 µg/ml propidium iodide (PI) for 10 min, washed, and finally, resuspended in 1 ml PBS.

A total of 10,000–20,000 events (cells) were collected, using an argon ion-based Coulter EPICS Elite analyzer operated at 15-mW electric power. The 90° light scatter (SS) was collected by a photomultiplyer tube by routing some of the scattered light through a 488-nm dichroic long-pass filter to a 488-nm band-pass filter. By using a 488-nm long-pass, laser-blocking filter and a 635-nm band-pass interference filter, PI fluorescence emission with a maximum at 620 nm was collected. Data acquisition and analysis were carried out using EPICS Elite Software (version 3.1).

Selective gating was employed on the FC data from infected and uninfected cultures to exclude undesirable fluorescence from cellular debris or doublets. The gating was accomplished by dual multiparameter analysis of the peak height of the PI signal and the PI fluorescence integral. Cell doublets with the same PI integral as intact single cells would show lower peak height signal and so would appear below the densely dotted area representing the single-cell population.

B. Assessment of the Level of Protein Expression

1. Biochemical Assessment of β–Galactosidase

The biochemical assay was an adapted version of the method proposed by Miller [8]. It is based on a colorimetric hydrolysis of the substrate O-nitrophenol-β-galactopyranoside (ONPG) by the β–galactosidase (β–gal); which frees dark yellow o-nitrophenol and galactose. As the enzyme β–galactosidase is an infection product that is released into the medium, while some remains within the cell, lysis of the cells was essential to measure the total β–galactosidase. Cell breakage was accomplished by repeated freezing and thawing. The biochemical assay involved loading of 20-µl aliquots of samples and standard β–galactosidase from *Escherichia coli* on a microwell plate at serial dilutions from 1:10 to 1:1280 to give a total volume of 200 µl/well. Next, the substrate ONPG was added at 20 µl in all wells after being prewarmed to 40°C for 1–2 min. The reaction was stopped with 50 µl of Na_2CO_3. The microwell plate was read in an enzyme-linked immunosorbent assay (ELISA) reader at 405 nm absorbance.

Absorbance readings of cell samples were interpolated on a standard curve, ranging between 0.1 and 12.5 U/ml. The interpolated value was multiplied by the dilution of the sample to give the β–galactosidase content in units per milliliter (U/ml).

2. Indirect Immunofluorescent Staining of β–Galactosidase

Monoclonal anti-β–gal antibody was used to bind the intracellular β–galactosidase which, in turn, was bound to a secondary polyclonal antibody, the antimouse IgG conjugated to fluoroscene isothiocyanate (FITC). After fixation, cells were washed twice in phosphate-buffered saline (PBS) + 1% NBS. An aliquot of 100 µl of monoclonal antibody diluted 1:100 in PBS was added to the cells for 30 min, washed twice in 0.5% Tween plus PBS, followed by the addition of antimouse·IgG–FITC conjugate (1:10) for 30 min. Two negative controls were used: unstained cells for background setting, and infected cells stained only with antimouse IgG–FITC conjugate. A protocol similar to the previously described PI-staining procedure was followed for dual DNA and β–gal detection. The FITC fluorescence emission was collected using a 550-nm short-pass dichroic filter and 525-nm band-pass filter.

III. RESULTS AND DISCUSSION

A. Insect Cell Cycle in a Batch Cultivation

There are fundamental differences between the cell cycles of mammalian and insect cells [4,6]. Typically, a mammalian cell culture contains a high fraction of G_1 cells during the whole culture period, so that the ratio of G_1 to G_2 cells is always higher than 1, reaching the highest value during the culture's stationary phase. The Sf9 insect cell cycle, however, is characterized by the apparent presence of a small fraction of G_1 cells as well as an additional small third peak comprising cells with tetraploid (8c) DNA content (Fig. 1). The presence of a mixture of diploid and tetraploid subpopulations has complicated the analysis of the cell cycle dynamics during batch cultivation of insect cells. It seems that the tetraploid population gradually increases at a very low rate over long-term passaging. In a recent analysis, an Sf9 sample obtained from the ATCC contained no polyploid cells. After what must be presumed to be a very large number of generations, the apparent ratio of diploid to tetraploid subpopulations is currently 2:1, although it varies between 1.6:1 in the lag phase and 2.3:1 in the midexponential phase of batch culture. In recent work in our laboratory, several diploid and tetraploid clones were isolated, and their cell cycle distributions were compared with those of the parent cell line during batch culture. In the parent cell line Sf9, the apparent ratio of the diploid G_1 (peak 1) to the diploid G_2/tetraploid G_1 (peak 2) cells changes with time during batch culture—being relatively higher in the exponential phase than in the stationary phase. In typical DNA distributions during batch culture (Fig. 2), after an initial increase (during the first 24 h of cell growth) in the fractions of diploid G_1 and S, the ratio of diploid G_1 to diploid G_2/tetraploid G_1 continuously decreased for the rest of the culture. Characteristically the diploid G_2/tetraploid G_1 peak became increasingly higher from the midlog growth phase and reached a maximum in the stationary phase. The diploid clones showed more similarity to the parent Sf9 cell line than the

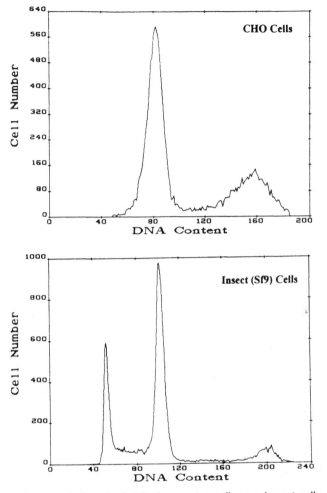

FIGURE 1 Cell cycle distribution—mammalian vs. insect cells.

tetraploid clones in their cell cycle kinetics. In the latter, the percentage of G_2 cells was relatively low during the early stages of growth, even though it was still distinctly higher than that normally found in mammalian cells. The presence of a high percentage of G_2 cells, specifically during the stationary phase, persuaded Fertig et al. [4] to suggest that the G_0 phase of insect cells is characterized by a 4c DNA content (G_2) and that cells, when exposed to stress, stay in this phase and do not proceed into G_1 phase.

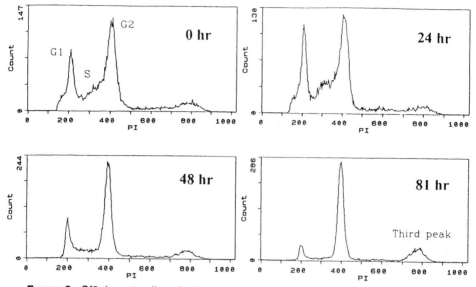

FIGURE 2 Sf9–insect cell cycle during a batch cultivation.

B. DNA and Orthogonal-Scattering Properties During the Infection Cycle

After infection of insect cells with wild-type baculovirus, an increase in the total DNA content is expected owing to the early viral multiplication (by 10–12 h post-infection virus particles begin to bud from the plasma membrane). The increase in DNA (measured by FC) can be compared with the percentage increase in the number of cells containing polyhedra (observed by light microscopy) to assess the accuracy and the advantages of the former method. In infected cells examined by electron microscopy, these polyhedra appeared as prominent, large, cuboidal crystals in the nucleus, together with large aggregates of fibrous structures.

During the post-infection period, the DNA content and distribution change considerably with time (Fig. 3). The increase in the viral DNA was indicated by the movement of cells from the G_1–G_2 into the higher than G_2 DNA content. Infecting cells at MOI 50 (MOI; *multiplicity of infection*, is defined as the number of virus particles per number of cells) led to an apparent increase in DNA content by about 20%, 24 h after infection. By 72 h, all infected cells show maximum DNA content and, after peaking, the DNA content starts to decrease owing to the increasing release of viral particles and the disintegration of cells.

Because of the DNA variations throughout infection, the percentage of cells with DNA higher than G_2 was selected as an indicator for infection. The

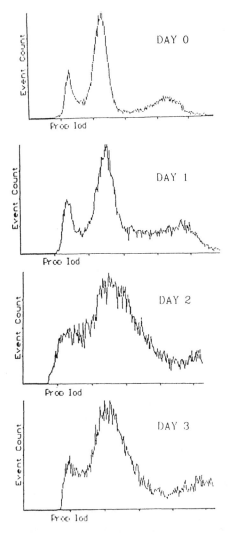

FIGURE 3 DNA distributions of an infected Sf9 insect cell culture with wild-type baculovirus (AcNPV).

data generated was further used to analyze the effect of MOI on infectivity. To exclude scale variabilities, data were normalized according to the following equation:

$$\left(\frac{(N_t - N_0)}{N_0}\right) \times 100 = X \tag{1}$$

N_t = the value of DNA at time t.
N_0 = the initial value of DNA at time zero.
X = the normalized percentage increase of the parameter at time t as related to the initial value at time zero.

The 90° light scatter was seen to increase strikingly with the progress of infection. The change of cell granularity, perhaps initially caused by free viral particle multiplication, but more noticeably later from polyhedra formation, was clearly reflected in the mean 90° light scatter. Data were again normalized using Eq. (1). These data were also used to identify differences in infectivity between cultures infected at different MOI.

1.   Effect of the Multiplicity of Infection on Polyhedra Formation

Insect cell cultures were infected with wild-type virus at MOI 1 and 50 in their late exponential phase. The rate of infection in the culture of MOI 50 was higher than in that of MOI 1. Both cultures, however, ended with similar maximum percentage infectivities at day 4 post-infection (Fig. 4a). The percentage infectivity was estimated as the proportion of cells that contained polyhedra to the total cell number.

When comparing the microscopic data with DNA FC analysis and measurements of 90° light scatter, it can be seen that the movement of cells from the "normal" cell cycle into higher levels of DNA, ahead of G_2 ploidy, increased substantially from day 1 up to day 3 post-infection and at a faster rate in the culture of MOI 50, compared with that of MOI 1 (see Fig. 4b). Such an increase could not be seen on the first day using light microscopy, as only a few polyhedra were apparent in the culture. The orthogonal-scattering properties of the infected cells showed a progress similar to the microscopically estimated percentage infectivity, although the difference between MOI 1 and 50 was more pronounced (see Fig. 4c).

2.   Detection of Infection in Cultures Infected at Different Growth Phases

The infection of cells from the stationary phase of batch culture with baculovirus caused a severe delay in the progress of infection and yielded a 12% lower infectivity than that obtained with cultures infected in the exponential phase (Fig. 5a). The foregoing effects were also reflected in the DNA multiplication seen ahead of the G_2 phase, this being 40% higher in the culture obtained from the exponential phase by day 1 postinfection (see Fig. 5b). The loss of DNA from the second day in this culture reflected the end of the infection and multiplication period and the start of the release of virus particles from infected cells. The DNA of the cells obtained from the stationary phase was also reduced from that of the second day, although polyhedral formation was still progressing. This was confirmed by the increase of 90° light scatter up to day 3 (see Fig. 5c).

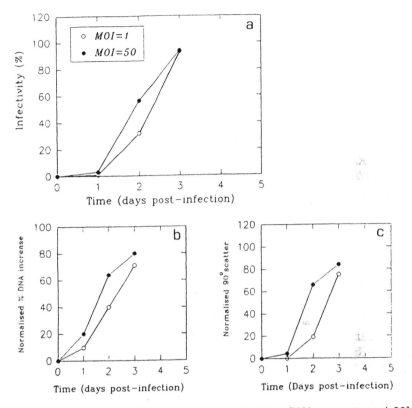

FIGURE 4 Comparison between percent infectivity, DNA contents and 90° light scatter in two cultures infected with the same wild-type virus (AcNPV) at different MOI (1 and 50).

Clearly, therefore, DNA monitoring by FC provides information on the level of infectivity long before this can be detected by microscopic means (i.e., by polyhedra detection). The results also show that the orthogonal properties of infected cells reflect precisely the increase in the rate of polyhedra formation. By using experimental data obtained from cultures infected at various MOI, the relation between percentage infectivity measured by microscopy and the 90° light scatter by FC (Fig. 6) could be assessed. A linear relation, with a correlation coefficient of 0.87, could be seen between the orthogonal scattering properties and the percentage of infected cells, as assessed by microscopy.

3. Infection Patterns Assessed by β-Galactosidase Formation

In a typical β–gal virus infection the product β–galactosidase is released extra-cellularly 1 day post-infection. A biochemical method [8] was used to monitor

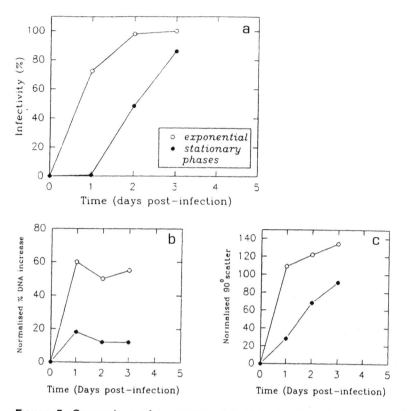

FIGURE 5 Comparison of percentage infectivity, DNA contents, and 90° light scatter in cultures infected with the same MOI (50) wild-type baculovirus (AcNPV) at different growth phases (stationary and exponential).

total and extracellular β–galactosidase from which intracellular amounts could be estimated. Patterns for β–galactosidase in an Sf9 culture infected with β–gal virus are given in Figure 7. Intracellular β–galactosidase was stained by an indirect immunofluorescent method as described in Materials and Methods (see Sec. II.B). The principle was based on a primary monoclonal antibody binding (anti-β–gal) to β–galactosidase and to which was sequenced a secondary fluorescent antibody (FITC–IgG polyclonal) that was detectable by FC.

Differences in β–galactosidase expression between the two cultures infected at MOI 1 and 50 were detected by both biochemical and FC techniques (Fig. 8). Although the rate of infection with the virus initially was faster in the culture of MOI 50, both cultures reached similar yields. The latter was also reflected in the extracellular viral titer measured by a time-consuming plaque

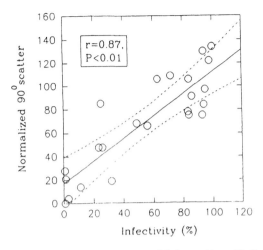

FIGURE 6 The relation of 90° light scatter with the percentage infectivity based on data from cultures infected with wild-type baculovirus (AcNPV) at various MOI.

assay (7-day duration). Although both techniques showed similarity in measuring intracellular β–galactosidase contents, FC has the advantages of being less tedious and time-consuming.

Recently, we have explored the feasibility of a rapid, quantitative FC assay of β–galactosidase with different lipophilic derivatives of the substrate fluorescein di-β–galactopyranoside (FDG). The fluorescent assay displayed much higher sensitivity to β–galactosidase activity compared with standard chromogenic assays, which made FC calibration problematic. Within a limit of substrate concentration of up to 33.3 μM of FDG (Molecular Probes, Inc., marketed as ImaGene) we have demonstrated that it is impossible to saturate β–galactosidase activity in infected cultures beyond 48 h post infection with MOI ranging from 0.1 to 10. We have also demonstrated that essentially a stoichiometric equilibrium exists between the hydrolyzed fluorescent substrate associated with viable cells and the hydrolyzed fluorescent substrate associated with the culture supernatant, and this made the resolution of a mixture of β–galactosidase-negative and β-gal-positive cells impossible.

IV. CONCLUSION

Flow cytometry can provide much information for the insect cell technologist more quickly and more reliably than other techniques. In this work, further demonstration has been given of the use of FC in process development for early and safe identification of the complex events occurring in the insect cell-

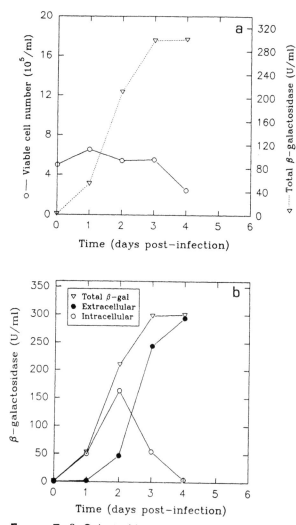

FIGURE 7 β–Galactosidase synthesis patterns in an Sf9 insect cell culture infected with recombinant (β–gal) virus at MOI 1. (a) Viable cell number and total (secreted and intracellular) β–galactosidase; (b) intracellular vs. extracellular β–galactosidase.

baculovirus system during the stages of growth, infection, and protein expression. Useful information can often be obtained up to 24 hs ahead of that available by more tedious and potentially unreliable assay methods.

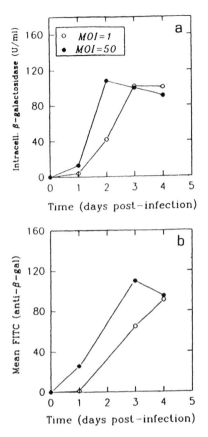

FIGURE 8 Comparison of β–galactosidase content between (a) biochemical assay [8] and (b) FC immunofluorescent method, in two cultures infected with β–gal virus at MOI 1 and 50.

REFERENCES

1. I. R. Cameron, R. D. Possee, and D. H. L. Bishop, Insect cell culture technology in baculovirus expression systems, *Trends Biotechnol. 7:*66 (1989).
2. M. Al-Rubeai, S. W. Oh, R. Musaheb, and A. N. Emery, Modified cellular metabolism in hybridomas subjected to hydrodynamic and other stresses, *Biotechnol. Lett. 12:*323 (1990).
3. J. L. Farmer, R. G. Hampton, and E. Boots, Flow cytometric assays for monitoring production of recombinant HIV-1g160 in insect cells infected with baculovirus expression vector, *J. Virol. Methods 26:*279 (1989).
4. G. Fertig, M. Kloppinger, and H. G. Miltenburger, Cell cycle kinetics of insect cell cultures compared with mammalian cell cultures, *Exp. Cell Res. 189:*208 (1990).

5. B. Schorpf, M. W. Howalt, and J. E. Bailey, DNA distribution and respiration activity of *Spodoptera frugiperda* population infected with wild-type and recombinant AcNPV, *J. Biotechnol. 15:*169 (1990).

6. M. Al-Rubeai, M. Kloppinger, G. Fertig, H. G. Miltenburger, and A. N. Emery, Monitoring of biosynthetic and biological activity in animal cell culture using flow cytometric methods, *Animal Cell Technology–Process and Products* (R. Spier et al., eds.), Butterworths-Heinemann, Oxford, 1992, p. 301.

7. M. D. Summers, and G. A. Smith, Manual of methods of for baculovirus vectors and insect cell culture procedures, *Tex. Agric. Exp. Stn. Bull. B1555:*1 (1987).

8. I. M. Miller, *Experiments in Molecular Genetics*, Cold Spring Harbor Laboratory, Cold Spring Harbor, NY, 1972.

13

Use of Protein Distribution to Analyze Budding Yeast Population Structure and Cell Cycle Progression

DANILO PORRO AND LILIA ALBERGHINA
Università degli Studi di Milano, Milan, Italy

I. INTRODUCTION

The understanding of a biological process is greatly increased by the availability of information on the physiological state of the growing biomass and by knowing how characteristics related to the physiological state are distributed within the cell population. In general, little information is available on this critical issue, since conventional sensors measure only average quantities in the population.

Flow cytometry is a powerful technology, the development of which started in the 1970s. It allows one to carry out measurements of physical or chemical bioparameters in the individual cell at rate so fast that the distribution of the characteristic in the whole population can also be accurately obtained in a very short time [1].

The application of flow cytometry to microbiology began in the late 1970s, the delay being due to the much higher sensitivity required for the analysis of microbial cells, when compared with mammalian cells. Studies in this area investigate both functional and structural properties. The applications range from the characterization of viable microbial cells, to the analysis of the DNA distribution; from the analysis of the dynamics of heterologous protein production, to the characterization of individual structural components; from the selection of high-producer cells, to the analysis of the cell protein distribution; from the study of the metabolic state, to the analysis of cell shape, and so forth. [1–10].

During the past several years, our laboratory has been engaged in developing flow cytometry as a tool to analyze budding yeast populations [reviewed in Ref. 11]. In fact, *Saccharomyces cerevisiae* is an important organism for the study of cell cycle control in eukaryotic cells [12,13]. In addition, it is a microorganism of wide biotechnological interest, since it is used for both conventional (e.g., biomass, ethanol, vitamins, or other.) [14] and advanced production (e.g., production of heterologous proteins, vaccines, chemicals, environmental remediation, and such.) [15–17]. Flow cytometric analyses allow a deeper understanding during both basic cell biology studies and biotechnological applications [11,18].

In this chapter, we discuss different technical approaches to the investigation of the protein distribution of exponentially growing and perturbed yeast populations which allow the determination of population structure(s) as well as the temporal and dynamic parameters of the cell cycle.

II. PROTEIN DISTRIBUTION ANALYSIS OF *SACCHAROMYCES CEREVISIAE* POPULATIONS

During the balanced exponential phase of growth (i.e., the growth phase in which all the bioparameters of the cellular population are maintained constant) [19], cell growth and nuclear division cycle events are coordinated by a threshold mechanism that activates the nuclear division cycle as described by Alberghina et al. [20,21]. Both daughter and parent cells express a critical cell size control during cell cycle progression, since they initiate DNA replication when they reach a critical protein threshold (called Ps) and enter into mitosis when they subsequently attain a second protein threshold (called Pm) [11,18,20–25]. The asymmetric division [26–34] that takes place in budding yeast yields quite complex population structures that are very sensitive to growth conditions [11,18,35–39]. In fact, the degree of size symmetry at division is growth rate-dependent, with more slowly growing cells dividing more asymmetrically [11, 27,28,31–34,38–40]. A further contribution to heterogeneity is made by the increase in size of the parent cells observed at each new generation [11,18,41]. Therefore, each population has a very complex structure, since older (and less frequent) parent cells have shorter generation times and produce large daughter cells, with shorter cycle times. The cell cycle of parents of different genealogical age and those of their daughters can be represented as indicated in Figure 1.

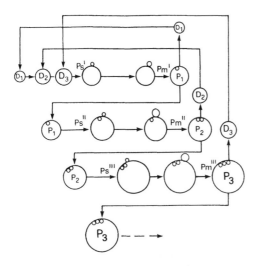

FIGURE 1 Model of the cell cycle of the yeast *S. cerevisiae*. All daughter cells begin to bud at the same relative protein threshold Ps^I, independently of the genealogical age of their parents. On the contrary, the critical protein threshold Ps required for parent cells to bud increases at each generation; thus, it has a value of Ps^{II} for the first generation, Ps^{III} for the second one, Ps^{IV} for the third generation, and so on. Since the budded period (Tb, the time necessary to growth from Ps to Pm) is of constant duration for all the genealogical ages [18], genealogically older parent cells and their daughters are larger at division. The attainment of the second protein threshold (Pm) is required to initiate the events of mitosis and cell division. Pm is also modulated by the genealogical age (Pm^I, Pm^{II}, Pm^{III} . . .).

Dynamic properties of both balanced and perturbed or transient states of growth can be investigated by extracting information from the protein distribution(s) by using appropriate mathematical models [11,42–47]. In fact, the protein distribution of a cell population represents the experimental parameter directly reflecting the cell population structure, which is strictly dependent on the growth laws of the cell population and on those of the single cell. During balanced exponential growth, the protein distribution is characteristic for each growth rate and quickly changes during growth transition or in perturbed conditions [11,18, 35–39,42].

Simulated and experimental protein distribution(s), derived for S228C yeast populations growing in batch cultures on YEP-glucose, YNB-glucose, and YNB-ethanol media are shown in Figure 2. The first quantitative information that can be obtained from a protein distribution is the average protein content per cell in the population ($\bar{P}$); $\bar{P}$ is related to the specific growth rate of the overall

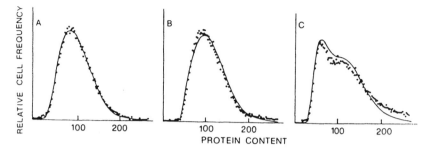

FIGURE 2 Experimental (dotted line) and simulated (continuous line) protein distributions of S288C *S. cerevisiae* cells exponentially growing in batch cultures on (A) YEP-glucose-, (B) YNB-glucose-, and (C) YNB-ethanol-based media. Exponentially growing cells were collected by centrifugation, sonicated, washed, and resuspended in 70% ethanol. The fixed cells are centrifuged, washed once with phosphate buffer (pH 7.4), then resuspended in 0.5 *M* sodium bicarbonate, containing 50 µg/ml of fluorescein isothiocyanate (FITC). After 30 min, cells are recovered by centrifugation and washed three times with phosphate buffer. The intensity of green fluorescence (> 520 nm) was determined on at least 100,000 cells with a fluorescence-activated cell sorter (FACS IV and/or FACStar^plus, Becton Dickinson), equipped with an argon ion laser yielding 200 mW at 488 nm. Simulated distributions have been obtained following a computer program previously described (see also Table 1). YEP; yeast extract (1% w/v), peptone (2% w/v); YNB; yeast nitrogen base (0.67% w/v); carbon source; 2% w/v or 2% v/v.

cell population, being higher at faster growth rates for both batch, continuous, and fed-batch growth conditions [11,18,27,35,38,46]. Mathematical analysis allows one to derive relevant parameters of a growing yeast population, such as the cycle time for daughter and parent cells of different genealogical age, and the proportion of each different subpopulation in the total population [11,42–47]. One can also determine the length of the budded phase (Tb, the time necessary for growth from Ps to Pm), which is the same for each subpopulation in any given growth condition [18]. Tables 1 and 2 summarize some experiments using yeast cells in batch growth on different substrates or at different dilution rates in glucose-limited chemostat fermentations.

    To relate changes in the population structure to environmental modifications (e.g., nutrient availability) and bioreactor dynamics (e.g., batch, continuous, and fed-batch growth), a mathematical model has been developed from the basic two-threshold cell cycle model, allowing prediction of the behavior of the growing population in the bioreactor during balanced as well as perturbed and transitory states [11,43,48–51].

    For budding yeast, the specific growth rate in batch, in chemostat, and in fed-batch conditions is a function of nutrient availability. Such availability is

TABLE 1 Structure of Yeast Populations Growing on Different Substrates (see also Fig. 2)

	T^a	$Tb^{b,c}$	$Tb^{b,d}$	Tp					Td				
				P.	1st	2nd	3rd	4th	D.	1st	2nd	3rd	4th
						(generations)					(generations)		
A	75	60	58	71(48)	77(27)	66(12)	62(6)	60(3)	84(52)	95(30)	75(13)	66(6)	62(3)
B	104	86	80	99(48)	107(27)	92(12)	86(6)	83(3)	117(52)	131(30)	104(13)	92(6)	87(3)
C	314	180	170	224(38)	253(19)	206(9)	187(6)	178(4)	476(62)	530(30)	449(16)	410(10)	393(6)

A,B,C refer to yeast cells exponentially growing on YEP-glucose, YNB-glucose, and YNB-ethanol, respectively.

[a] T, generation time of the overall population (min).

[b] Tb, length of the budded phase (min).

[c] As determined following protein distribution deconvolution.

[d] As determined using the following equation $Tb:\log_2 (1 + Fb)T$ [39], where Fb is the fraction of the budded cells in the whole population and T is the duplication time of the overall population.

Tp, Td, generation time (min) of the whole parent and daughter populations (P., D.), as well as of genealogical aged (1st to 4th) parent and daughter subpopulations, respectively. Data in parentheses represent the relative cell frequency.

TABLE 2 Structure of Yeast Populations Growing at Different Dilution Rates in Glucose-Limited Chemostat Fermentations

D^a	$Fb(\%)^b$	Tb (min)^c	Tp (min)^c	Td (min)^c
0.07(591)	25	191	295	1384
0.20(207)	65	152	193	300
0.29(143)	66	105	131	172
0.31(134)	66	98	123	169
0.35(118)	64	85	107	145
0.40(103)	67	78	99	147
0.48(86)	75	70	86	94

[a]D, Dilution rate (h^{-1}). Data in parentheses report the generation times (T) of the yeast population, (min).
[b]Fb, Fraction of budded cells as determined by microscopic counting.
[c]Tb, Tp, and Td represent the length of the budded period, the generation time of the parent cell population, and the generation time of the daughter cell population, respectively, as calculated from the model.

directly dependent on the amount of nutrient at the beginning of the fermentation, on the dilution rate, or on the controlled feeding rate, respectively, and it determines the setting for the two-threshold protein contents Ps and Pm. Thus, the model basically takes into account the relation between the substrate concentration and the setting of Ps and Pm. The more interesting applications of this model are related to the study of perturbed and transitory states. An example, showing the addition of glucose to a glucose-limited chemostat culture, is given in Figure 3. In a first phase, the increased availability of glucose yields a shift to the right of the protein distribution, with a temporary decrease of the cell number concentration. These results can be interpreted by considering that the increase of glucose concentration increases both Ps and Pm. The increase of Pm delays cell division, thus explaining the initial cell number reduction and the shift to the right (i.e., to higher values) of the protein distribution. After exhaustion of glucose, a cohort of cells appears on the left (i.e., cells with a smaller size) of the distribution. This cohort of cells represents the newborn daughter cells generated from parent cells that budded before the critical level of Ps came down to its new steady-state value and that divide at the new lower steady-state level of Pm [38]. The simulated transitory states are very close to the experimental findings (see Fig. 3) [45, 48–50].

This kind of analysis is very useful; for instance it helped us gain insight on the effects on Ps, Pm, and, therefore, on the population structure after dilution rate step-up [48–50] and to understand the mechanism inducing and maintaining a partial cell synchronism at very low-dilution rates in chemostat cultures [36,37,

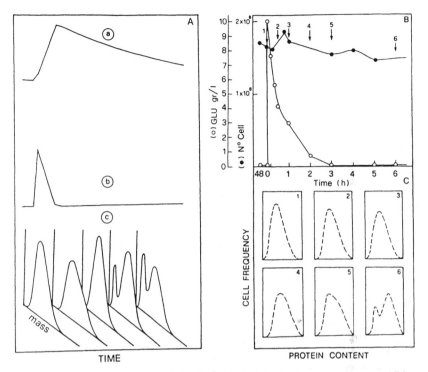

FIGURE 3 Experimental and simulated behaviors after a glucose addition to steady-state, glucose-limited chemostat culture. (A) Simulated data: Time course of cell number concentration (a), substrate (b, glucose) and mass distribution (c, protein distribution). (B,C) Experimental data: (B) Time course of cell number concentration (solid circles), glucose concentration (open circles), and (C) protein distributions (1-6, see arrows in B).

51]. It also helped us relate the dynamics of the cell population to the control of feeding rate in fed-batch processes [35] and to characterize, in a quantitative way, the dynamics of cell lysis induced in the cell parent population by the overexpression of the transcriptional activator *GAL4* [11,52].

III. PROTEIN DISTRIBUTION ANALYSIS DURING PROGRESSION OF THE CELL CYCLE

The main assumptions at the basis of the two-threshold cell cycle model are that (a) single-cell mass dynamic follows an exponential function; and (b) all individual cells grow with the same first-order rate constant, independently of their genealogical age (i.e., independently of their cell cycle length) and, during balanced growth, such a constant value is equal to the specific growth rate of the overall population.

Such assumptions have not been tested experimentally in unperturbed, exponentially growing population, since direct determinations are not available, but they have been calculated by inference from other data, or estimated following synchronization procedures [40]. However, synchronization procedures are likely to perturb the cell population dynamics, and the results obtained may not be an accurate representation of normal growth behavior.

A new flow cytometric procedure that allows the determination of dynamic properties of asynchronously growing *S. cerevisiae* cells has recently been developed [53]. The procedure is based on labeling the cell wall with a lectin, such as concanavalin A (ConA), conjugated to a fluorescent marker, such as fluorescein isothiocyanate (FITC). Cells were harvested in the exponential growth phase, stained with ConA–FITC and, then, allowed to grow in the same culture medium. The biosynthesis of the cell wall is determined by preexisting cell wall structures and environmental conditions, in addition to the genetic background [54]. After a yeast cell has divided, the daughter cell is growing in an apolar way and continues its diffuse expansion until it switches to bud growth. The parent cell, with one newly acquired bud scar, also switches to bud growth, but after a much shorter time. Localization of cell wall synthesis during growth of the bud has been demonstrated with fluorescent dyes [55,56]. In general, the cell wall of the growing bud is not derived from the material of the parent cell wall, but rather, it is synthesized de novo [54–56]. Thus, after staining the cells, resuspension, and the following new growth, newly synthesized cell wall is not stained, whereas the older cell wall components retain the initial fluorescence [53]. Staining conditions have been achieved that do not perturb cell growth after resuspension of stained cells in the new growth medium [53].

The analysis of the staining patterns has been used to provide direct information on the properties of the growing cells, such as the specific growth rate of the overall population, duration of the budded phase (Tb), and kinetic pattern of growth [53].

The procedure previously described has been further improved by the contemporary determination of the experimental parameter directly reflecting the cell population structure (i.e., the cell size). Cell size has been estimated from the single-cell protein content, as assessed by staining with tetramethylrhodamine isothiocyanate (TRITC). The double-staining procedure allows one to gain quantitative information on the dynamic(s) of growth, as indicated in Figure 4. In fact, at the time of staining with ConA–FITC, all the cell walls are completely stained (region 3 in the figure). Since the cells occupy different cell cycle positions, they will generate daughters with a gradually decreasing degree of surface stain. The partially stained cells (region 2) represent newborn daughter cells coming from cells stained during the budded phase $(S+G_2+M+G_1{}^*)$ of the

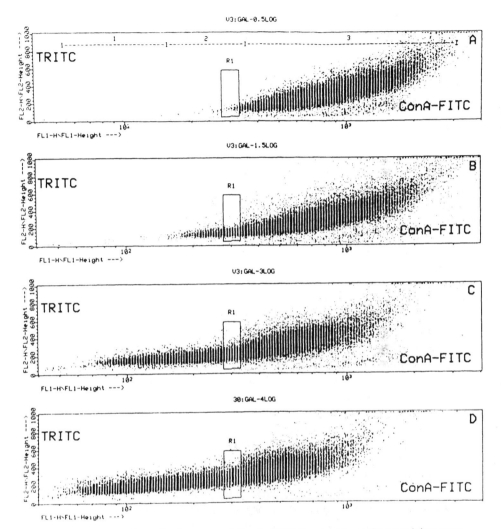

FIGURE 4 Examples of cytograms of ConA–FITC fluorescence (log scale) versus TRITC signal (linear scale) taken at different times after cell wall staining and following resuspension. X4004 yeast cells growing exponentially on a YNB–galactose-based medium were collected by centrifugation and resuspended in precooled fresh YNB-based medium, containing 120 µg/ml of conjugated ConA–FITC (approx. 3.6 mol of FITC per mol of lectin; SIGMA, St. Louis, MO) at a cell concentration of 2 × 10⁸/ml. After 7 mins of staining, cells were recovered by rapid centrifugation and resuspended in an identical preheated fresh medium (2 × 10⁶/ml). All the operations were carried out at 4°C. At different times after resuspension and new growth, yeast cells were collected, fixed in ethanol (70%) for 20

cell cycle, whereas unstained cells (region 1) represent the newborn daughters from cells stained in the unbudded phase (G_1). Since each cell retains the same amount of fluorescence (i.e., the dye attached to the cells is diluted only by the new growth), the selection of a cohort of newborn daughter cells of the same fluorescence (gate R1 in the figure) in the growing population allows us to follow their size increase, as indicated by the TRITC signal, during the cell cycle progression. An example showing the time course changes in the patterns of ConA–FITC versus TRITC and the method of selection of such a cohort is presented in Figure 4. The time courses of the mean cell size, relative CV (coefficient of variation), and fraction of selected cells are shown in Figure 5 for cohort of partially stained daughter cells selected from a population in balanced exponential growth on YNB-galactose.

The very strong agreement between the specific growth rate value of the overall population ($0.215 \ h^{-1}$), as determined by the increase of cell number concentration, and the specific growth rate value of the daughter cell size increase ($0.213 \ h^{-1}$), as determined by flow cytometry, strongly validates the procedure employed. In fact, for an exponentially growing population the rate of increase in cell number concentration must be identical with the rate of increase of the cell size throughout the cell cycle progression.

More importantly, Figure 5 shows that advanced instrumentation and measuring techniques are able to provide direct information on the dynamic properties of individual cells, such as (a) the dynamics of cell size increase (shown to fit accurately an exponential function), (b) the specific growth rate of this increase and (c) the cell age at specific setpoints of the cell cycle.

Furthermore, data, such as those shown in Figure 5 provide a sensitive approach to the estimation of the daughter cell cycle duration. Figure 5 shows that an exponential increase of the cell size (with a coefficient of correlation higher than 0.99) has been observed for 4.5 h, which is a duration much longer than the duplication time of the overall population (3.2 h). After this time (i.e., the switch from closed to open circles in the figure), the exponential order rate strongly decreases. This decrease has been related to the cellular division; interestingly, such a time deviation corresponds to the calculated cell cycle length for the

---

(FIGURE 4 continued) min (4°C), washed, and the total cell protein stained with TRITC following the procedure described in Figure 1. At time $T = 0$ the cell wall of each cell is completely stained (see A, region 3); the cells stained in the budded phase will generate partially stained daughter cells (region 2), whereas cells stained in the unbudded phase will generate daughter unstained cells (region 1; see also text). R1 represents the gate used on the different cytograms for the selection of newborn partially stained daughters cells of the same age (i.e., born at the same time). A, B, C, and D refer to yeast samples withdrawn at time $T = 0.5$, $T = 1.5$, $T = 3$, and $T = 4$ h after resuspension, respectively.

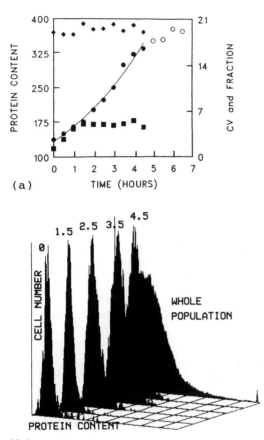

(a)

(b)

FIGURE 5 Cell size behavior during the cell cycle progression of newborn partially stained daughter cells growing, after staining and resuspension ($T = 0$), in YNB-gal-based medium

(A) (●, ○) Average cell size (expressed as average channel number of the relative protein distribution); (◆) coefficient of variation of the relative protein distribution; (■) fraction of cells/gate. (B) Cell protein distribution behavior during the cell cycle progression of daughter cells: 0, protein distribution of newborn daughter cells; 1.5, 2.5, 3.5, and 4.5, protein distributions observed after 1.5, 2.5, 3.5, and 4.5 h of growth (i.e., dividing daughter cells). The last histogram represents the protein distribution of the whole population.

daughter cell population (4.41 and 2.51 h for the daughter and parent cell sub-populations, respectively). Such agreement between estimated and calculated cell cycle duration has been observed for populations growing with different

TABLE 3 Comparison Between Growth Rate of the Cell Size Increase of the Daughter Cells and Growth Rate of the Whole Population[a]

	Specific growth rate (h^{-1})	
Experiment	Cell size of the daughter cells	Cell number concentration[b]
A	0.213 (0.99)	0.215 (0.99)
B	0.188 (0.99)	0.182 (0.99)
C	0.207 (0.99)	0.211 (0.99)
D	0.226 (0.99)	0.224 (0.99)
E	0.251 (0.98)	0.250 (0.99)
F	0.225 (0.99)	0.211 (0.99)
G	0.206 (0.99)	0.215 (0.99)
H	0.324 (0.98)	0.327 (0.99)

[a]The data relative to the daughter cell population represent the estimated rate of the exponential increase in size through the cell cycle progression, whereas data relative to the whole population represent the calculated growth rate of the overall population as determined by cell number increase (data in parentheses represent the relative first-order coefficient of correlation).
[b]For the determination of the cell number concentration a Coulter Counter Channalyzer was used.

overall growth rates, reflecting specific different cell cycle lengths (data not shown).

Cells growing with different overall growth rates and double-stained with the procedure described were analyzed; Table 3 summarizes some of these experiments.

IV. CONCLUSIONS AND PERSPECTIVES

The protein distribution is the parameter directly related to the population structure, and it is strongly dependent on the laws of growth of both the single cell and those of the population.

A deeper understanding of the population dynamics may be obtained by extracting information from such experimental data by the availability of appropriate algorithms. Several examples described in this review, and in preceeding papers indicate that a quite exhaustive analysis can be obtained. For instance, for both balanced-and perturbed-growing populations, the length of the cell cycle for cells with different genealogical ages, the relative cell frequencies, the length of the G_1 cell cycle phase, and the length of the budded phase, all can be determined, as well as many other parameters.

A further development of the experimental analysis is available in a procedure that allows us to tag the cells in a specific cell cycle stage with an appropriate fluorochrome and then follow the dynamics of growth with a second fluorochrome. The example described in this chapter indicates the potential of such an approach. The more interesting data obtained in this way are the specific growth rate at the single-cell level as well as the cell age at a single setpoint through the cell cycle, the cell size of the newborn as well as that of the dividing daughter cells and the duration of the daughter cell cycle. Integration between the approaches has been also discussed.

This kind of investigation, once again, indicates the potential of flow cytometry for cell cycle studies. Moreover, biotechnological production processes can be analyzed by the procedures described. For instance, in many biotechnological processes, production may be influenced by the age-dependent physiological and biochemical heterogeneity of the population. The quantitative analyses of protein distributions and their combination with other flow cytometric determinations should allow us to better describe, monitor, and in the final consideration, validate the process.

REFERENCES

1. H. M. Shapiro, *Practical Flow Cytometry*, Alan R. Liss, New York, 1988.
2. J. E. Bailey, D. W. Agar, and M. A. Hjiorsto, Acquisition and interpretation of flow microfluorimetry data on microbial populations, *Computer Application in Fermentation Technology*, Society of Chemical Industry, 1982, p. 157.
3. J. E. Bailey, J. Fazel-Madjlessim, D. N. Mcquitty, L. Y. Lee, and J. A. Oro, Measurement of structured microbial population dynamics by microfluorimetry, *AIChE J.* 24:570 (1978).
4. B. S. Dien and F. Srienc, Bromodeoxyuridine labeling and flow cytometric identification of replicating *Saccharomyces cerevisiae* cells: lengths of cell cycle phases and population variability at specific cell cycle positions, *Biotechnol. Prog.* 7:291 (1991).
5. P. D. Eitzman, J. L. Hendrick, and F. Srienc, Quantitative immunofluorescence in single *Saccharomyces cerevisiae* cells, *Cytometry* 10:475 (1989).
6. Y. Nishimura and J. E. Bailey, Bacterial population dynamics in batch and continuous flow microbial reactors, *AIChE J.* 27:73 (1980).
7. E. Sahar, R. Nir, T. Molcho, and R. Lamed, Flow cytometry of micro-colonies as an approach for mutant selection and industrial strain improvement, Proc. 6th European Congress on Biotechnology, Vol 1, Florence, Italy (L. Alberghina, L. Frontali, and P. Sensi, eds.), Elsevier Science, New York, 1993, p. 571.
8. J. M. De La Fluente, A. Alvarez, C. Nombela, and M. Sanchez, Flow cytometric analysis of *Saccharomyces cerevisiae* autolytic mutants and protoplasts, *Yeast* 8:39 (1993).

9. D. Porro, C. Smeraldi, E. Martegani, B. M. Ranzi, and L. Alberghina, Flow cytometric determination of the respiratory activity in growing *Saccharomyces cerevisiae* populations, *Biotechnol, Prog. 10*:193 (1994).

10. G. B. J. Dubelaar, J. W. M. Visser, and M. Donze, Anomalous behaviour of forward and perpendicular light scattering of a cyanobacterium owing to gas vacuoles, *Cytometry 8*:405 (1987).

11. L. Alberghina and D. Porro, Quantitative flow cytometry: analysis of protein distributions in budding yeast. A mini review, *Yeast 9*:815 (1993).

12. K. Nasmyth, Control of the yeast cell cycle by the Cdc28 protein kinase, *Curr. Opin. Cell Biol. 5*:166 (1993).

13. A. B. Futcher, Yeast cell cycle, *Curr. Opin. Cell Biol. 2*:246 (1990).

14. A. Fiechter, G. F. Fuhrmann, and O. Kappeli, Regulation of glucose metabolism in growing yeast cells, *Adv. Microb. Physiol. 22*:2031 (1981).

15. M. A. Ramonos, C. A. Scorer, and J. J. Clare, Foreign gene expression in yeast: a review, *Yeast 8*:423 (1992).

16. R. G. Buckholz, Yeast system for expression of heterologous gene products, *Curr. Opin. Biotechnol. 4*:538 (1993).

17. D. Porro, E. Martegani, B. M. Ranzi, and L. Alberghina, Lactose/whey utilization and ethanol production by transformed *Saccharomyces cerevisiae* cells, *Biotechnol. Bioeng. 39*:799 (1992).

18. M. Vanoni, M. Vai, L. Popolo, and L. Alberghina, Structural heterogeneity in populations of budding yeast *Saccharomyces cerevisiae*, *J. Bacteriol. 156*:1282 (1983).

19. P. Jagers, Balanced exponential growth: what does it mean and when is it there? *Biomathematics and Cell Kinetics* (A. J. Valleron and P. D. M. McDonal, eds.), Elsevier North-Holland, Amsterdam, 1978, p.21.

20. L. Alberghina, Dynamics of the cell cycle in mammalian cells, *J. Theor. Biol. 69*:633 (1977).

21. L. Alberghina and L. Mariani, Control of cell growth and division. *Biomathematics and Cell Kinetics* (A. J. Valleron and MacDonald, eds.), Elsevier North-Holland, Amsterdam, 1978, p. 89.

22. L. Alberghina, E. Martegani, L. Mariani, and G. Bortolan, A bimolecular mechanism for cell size control of the cell cycle, *Biosystems 16*:297 (1984).

23. L. Alberghina, Control and variability in the cell cycle, *Progress in Cell Cycle Controls*, 6th European Cell Cycle Control Workshop (J. Chaloupka, A. Kotyk, and E. Streiblova, eds.), Czechoslovak Academy of Sciences, 1983, p. 12.

24. L. Alberghina, L. Mariani, and E. Martegani, Analysis of a model of cell cycle in eukaryotes, *J. Theor. Biol. 87*:171 (1980).

25. L. Alberghina, L. Mariani, and E. Martegani, Cell cycle modelling, *Biosystems 19*:23 (1986).

26. L. H. Hartwell, *Saccharomyces cerevisiae* cell cycle, *Bacteriol. Rev. 38*:164 (1974).

27. G. C. Johnston, J. R. Pringle, and L. H. Hartwell, Coordination of growth and cell division in the yeast *Saccharomyces cerevisiae*, *Exp. Cell Res. 105*:79 (1977).

28. G. C. Johnston, R. A. Singer, S. O. Sharrow, and M. L. Slater, Cell division in the yeast *Saccharomyces cerevisiae* growing at different rates, *J. Gen. Microbiol. 118*:479 (1980).

29. J. R. Pringle and L. H. Hartwell, The *Saccharomyces cerevisiae* cell cycle,. *Cold Spring Harbor Monogr. Ser.* 1981, p.97.

30. L. Alberghina, L. Mariani, and E. Martegani, Modelling controls and variability of the cell cycle, *Mathematics in Biology and Medicine* (V. Capasso, E. Grosso, and S. L. Paveri-Fontana, eds.), Springer-Verlag, New York, 1983, p. 239.

31. M. L. Slater, S. O. Sharrow, and J. J. Gart, Cell cycle of *Saccharomyces cerevisiae* in populations growing at different rates, *Proc. Natl. Acad. Sci. USA 74*:3850 (1977).

32. B. L. A. Carter, The control of cell division in *Saccharomyces cerevisiae*, *The Cell Cycle* (P. C. L. John, ed), Cambridge University Press, Cambridge, 1981, p. 99.

33. B. L. A. Carter and M. N. Jagadish, Control of cell division in the yeast *Saccharomyces cerevisiae* cultured at different growth rates, *Exp. Cell Res. 112*:373 (1978).

34. C. B. Tyson, P. G. Lord, and A. E. Wheals, Dependency of size of *Saccharomyces cerevisiae* on growth rate, *J. Bacteriol. 138*:92 (1979).

35. L. Alberghina, B. M. Ranzi, D. Porro, and E. Martegani, Flow cytometry and cell cycle kinetics in continuous and fed-batch fermentations of budding yeast, *Biotechnol. Prog. 7*:299 (1991).

36. E. Martegani, D. Porro, B. M. Ranzi, and L. Alberghina, Involvement of a cell size control mechanism in the induction and maintenance of oscillations in continuous cultures of budding yeast, *Biotechnol. Bioeng. 36*:453 (1990).

37. D. Porro, E. Martegani, B. M. Ranzi, and L. Alberghina, Oscillations in continuous cultures of budding yeast: a segregated parameter analysis, *Biotechnol. Bioeng. 32*:411 (1988).

38. B. M. Ranzi, C. Compagno, and E. Martegani, Analysis of protein and cell volume distribution in glucose-limited continuous cultures of budding yeast, *Biotechnol. Bioeng. 28*:185 (1986).

39. L. Alberghina, L. Mariani, E. Martegani, and M. Vanoni, Analysis of protein distribution in budding yeast, *Biotechnol. Bioeng. 25*:1295 (1983).

40. C. L. Woldringh, P. G. Huls, and N. O. Vischer, Volume growth of daughter and parent cells during the cell cycle of *Saccharomyces cereviasiae* a/alpha as determined by imagine cytometry, *J. Bacteriol. 175*:3174 (1993).

41. G. C. Johnston, C. W. Ehrhardt, A. Lorincz, and B. L. A. Carter, Regulation of cell size in the yeast *Saccharomyces cerevisiae*, *J. Bacteriol. 137*:1 (1979).

42. L. Alberghina, E. Martegani, and L. Mariani, Analysis of protein distribution in populations of budding yeast based on a structured model of cell growth, *Modelling and Control of Biotechnical Processes* (A. Halme, ed.), Pergamon Press, New York, 1983, p. 83.

43. L. Cazzador and L. Mariani, Structured modelling and parameter identification of budding yeast populations, *Computer Applications in Fermentation Technology* (N. M. Fish, R. I. Fox, and N. F. Thornhill, eds.), Elsevier, New York, 1987, p. 211.

44. L. Mariani, L. Alberghina, and E. Martegani, Mathematical modelling of cell growth and proliferation, *IFAC Modelling and Control of Biomedical System* (C. Cobelli and L. Mariami, eds.), Pergamon Press, New York 1988, p. 269.

45. L. Mariani, E. Martegani, and L. Alberghina, Yeast population models for monitoring and control of biotechnical processes, *IEE Proc. 133*:210 (1986).

46. E. Martegani, L. Mariani, and L. Alberghina, Yeast biotechnological process monitored by analysis of segregated data with structured models, *Modelling and Control of Biotechnical Processes* (A. Johnston, ed.), Pergamon Press, Oxford, 1985, p. 237.

47. E. Martegani, M. Vanoni, and D. Delia, A computer algorithm for the analysis of protein distribution in budding yeast, *Cytometry 5*:81 (1984).

48. L. Cazzador, L. Alberghina, E. Martegani, and L. Mariani, Bioreactor control and modeling: a simulation program based on a structured population model of budding yeast, *Bioreactors and Biotrasformation* (G. W. Moody and P. B. Baker, eds.), Elsevier, New York 1987, p. 64.

49. L. Mariani, L. Cazzador, E. Martegani, and L. Alberghina, A bioreactor model and a simulation program based on a structured population model of budding yeasts, Proc. 4th European Congress on Biotechnology, vol.3, Amsterdam, Netherlands, (O. M. Neijssel, R. R. van der Meer, and K. C. A. M. Luyben, eds.), Elsevier Science Publisher, Amsterdam, 1987, p. 118.

50. L. Alberghina, E. Martegani, L. Mariani, and L. Cazzador, Flow cytometry for monitoring and controlling budding yeast fermentations. Proc. 4th European Congress on Biotechnology, vol. 4, Amsterdam, Netherlands, (O. M. Neijssel, R. R. van der Meer, and K. C. A. M. Luyben, eds.), Elsevier Science Publisher, Amsterdam, 1987, p. 467.

51. L. Cazzador, L. Mariani, E. Martegani, and L. Alberghina, Structured segregated models and analysis of self-oscillating continuous cultures, *Bioprocess Eng. 5*:175 (1990).

52. E. Martegani, L. Brambilla, D. Porro, B. M. Ranzi, and L. Alberghina, Alteration of cell population structure due to cell lysis in *Saccharomyces cerevisiae* cells overexpressing the *GAL4* gene, *Yeast 9*:575 (1993).

53. D. Porro and F. Srienc, Tracking of individual cell cohorts in asynchronous *Saccharomyces cerevisiae* populations, *Biotechnol. Prog. 11*:342–347 (1995).

54. R. Schekman and P. Novick, The secretory process and yeast-cell surface assembly, The molecular biology of the yeast *Saccharomyces*, *Metabolism and Gene Expression* (J. N. Strathern, E. W. Jones, and J. R. Broach, eds.), Cold Spring Harbor Laboratory, Cold Spring Harbor, NY, 1982, p. 361.

55. J. S. Tkacz, E. B. Cybulska, and J. O. Lampen, Specific staining of wall mannan in yeast cells with fluorescein-conjugated concanavalin A, *J. Bacteriol. 105*:1 (1971).

56. K. L. Chung, R. Z. Hawirko, and K. Isaac, Cell wall replication in *Saccharomyces cerevisiae*, *Can. J. Microbiol. 11*:953 (1965).

14

Cell Cycle Analysis
of Microorganisms

KIRSTEN SKARSTAD, ROLF BERNANDER, STURE WOLD,
HARALD B. STEEN, AND ERIK BOYE
Institute for Cancer Research, Oslo, Norway

I. INTRODUCTION

Detailed analyses of the bacterial and yeast cell cycles are now possible with
flow cytometry. We will focus here on fundamental aspects of cell cycle analysis
of the bacterium *Escherichia coli*, and demonstrate how DNA and cell size
distributions may be used to calculate key parameters of the bacterial cell cycle.
Also when flow cytometry is merely used to monitor the state of a microbial
culture, knowledge of basic DNA replication and cell division patterns is useful,
since the DNA and size distributions are very sensitive to changes in the growth
conditions. [For reviews on basic cell cycle analysis and advances of flow
cytometry in applied microbiology the reader is referred to Ref. 1 and 2, respec-
tively].

Even though applied to microbiology from the very beginning [3,4], most
flow cytometric applications have been developed for mammalian cells, a fact
that is reflected in the design of the instruments on the market. The volume of a
bacterium is three orders of magnitude smaller than that of a mammalian cell,
and so are the amounts of the various intracellular macromolecules. Because of
the small size and low DNA content, instrument sensitivity is critical. This

chapter, therefore, also contains a description of the microscope-based flow cytometer that has been developed in our laboratory and that is especially well suited for microbial measurements.

II. THE BACTERIAL CELL CYCLE

Escherichia coli, the most extensively studied of all bacteria, contains a single circular chromosome that is replicated bidirectionally from a unique origin of replication. By varying the nutrient availability in the medium, the bacteria may be grown at a wide range of growth rates, accompanied by large variations in cell size and DNA content. This is in contrast with mammalian cells, for which the DNA content of exponentially growing cells of a given type usually vary only between diploid and tetraploid. In slowly growing *E. coli* cells, with generation times of several hours, the cell cycle appears similar to that of higher eukaryotes [5,6]. The prereplication, replication, and postreplication periods are termed B, C, and D, respectively, and may be compared with the G_1, S, and G_2 phases of the eukaryotic cell cycle. DNA distributions of slowly growing bacteria in exponential phase exhibit a peak at one genome, corresponding to the B-period cells, a ridge between one and two genomes, corresponding to the replicating C-period cells, and a peak at two genomes representing D-period cells [5,6]. As in DNA histograms from higher eukaryotes, a histogram of slowly growing bacteria may be used to directly determine the fraction of cells in the different periods of the cell cycle. By also taking into consideration the exponential age distribution (that there are twice as many newborn cells as dividing cells) and the generation time of the culture, the duration of the different cell cycle periods may be calculated.

When *E. coli* cells are grown in media supporting faster growth, it takes 40–60 min to replicate the chromosome (the C period), depending somewhat on the strain [6–8], and the D period lasts 20–30 min. The time required from initiation of replication to cell division, therefore, is 60–90 min. The generation time of an *E. coli* culture grown in rich medium, however, is much shorter, about 20 min. This means that replication must be initiated in the generation before, at two replication origins in the mother cell, or even two generations before, in the grandmother cell at four origins, to have time enough for the C and D periods. This is the Cooper-Helmstetter model for replication in rapidly growing bacteria [7] and is depicted in Figure 1 for two different growth rates; 60- and 25-min– generation times. For the sake of clarity, we assume that the C period lasts exactly 40 min and the D period 20 min in both examples. The cells in the 60-min culture will have no B period because the entire 60 min are required to go through the C and D periods. Therefore, initiation of replication must occur when the cell is newborn (see Fig. 1B). Replication proceeds bidirectionally and terminates after 40 min. The cells then contain two fully replicated chromo-

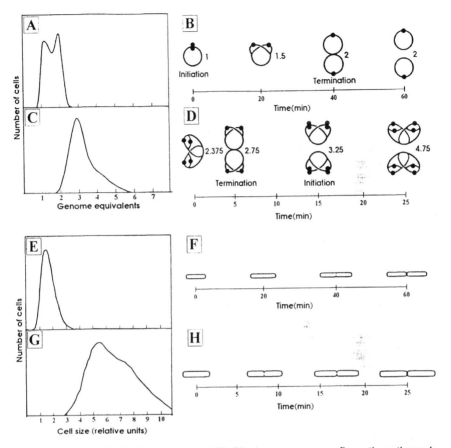

FIGURE 1 (A, C) DNA histograms and (B, D) chromosome configurations through the cell cycle of *E. coli* cells with a C period of 40 min and a D period of 20 min. (A, B): With a generation time of 60 min; (C, D) with a generation time of 25 min. Large circular structures represent genomes and small, filled circles represent replication origins. The DNA histograms were generated with the aid of a computer simulation routine [6] assuming 10% biological variation and 10% variation caused by staining and measuring variability. The histograms closely resemble experimental histograms with the given cell cycle parameters. (E, G) Light-scattering histograms were experimentally obtained from cultures growing with generation times of approximately 60 and 25 min. (F, H) The theoretical cell dimensions through the division cycle were calculated from Ref. 25.

somes throughout the 20-min D period. This replication pattern is reflected in the DNA histogram (see Fig. 1A). The DNA histogram is composed of an array of "channels," each representing a particular DNA content. The number of cells with a certain DNA content is dependent on the rate of DNA synthesis. The D-period cells have no DNA synthesis; thus, all accumulate at the same DNA content, namely two genome equivalents. The newborn cells initiate replication at one origin, and the rate of DNA synthesis in these cells is one genome per 40 min. Thus, all the C-period cells have the same rate of DNA synthesis; accordingly, the cells "travel through" the channels representing DNA contents from one to two genome equivalents at a constant rate. This would have led to the same number of cells in each channel if it had not been for the exponential age distribution. Because the age distribution is also reflected in the histogram, there are more young cells with low DNA contents, than older cells with more DNA. The DNA histogram of the 60-min culture is thus composed of C-period cells, giving rise to an exponentially declining distribution stretching from one to two genome equivalents, and D-period cells, which give a peak at two genome equivalents. Because of biological as well as methodological variability, the histogram does not appear as a sharp exponential function plus a sharp peak at two genome equivalents. The degree to which the "theoretical" shape is obscured is determined by the variation in the cell cycle parameters between the individual cells in the culture, and by the variability in the staining and flow cytometric measurements, the latter being of the order of a few percent.

The replication pattern of cells growing with a generation time of 25 min is more complex (see Fig. 1D). Initiation of replication occurs 10 min before cell division in the grandmother cell, at four replication origins. At cell division, these four pairs of replication forks have proceeded one-fourth of the way to the termini (one-fourth, because 10 min of the 40-min replication period has passed). The two pairs of replication forks from the previous round of initiations have proceeded seven-eighths of the way to the termini and will terminate 5 min after cell division. The DNA contents of newborn and dividing cells are therefore 2.375 and 4.75 genome equivalents, respectively. This is reflected in the DNA histogram (see Fig. 1C), which stretches from a little more than two to about five genome equivalents. The shape of the histogram again reflects the rate of DNA synthesis and the age distribution. There are two points in the cell cycle at which the rate of DNA synthesis changes, one is at initiation of replication and the other is at termination of replication. Five minutes after a cell is newborn, two of the ongoing six forks finish replication and, therefore, the rate of DNA synthesis is reduced by one-third (see Fig. 1D). At 15 min, initiation occurs at four origins and the rate of DNA replication increases threefold. This large change in the rate of DNA synthesis (at about 3.25 genome equivalents) is noticeable in the DNA histogram, in which the number of cells per channel is sharply reduced because of the increased rate of DNA synthesis (see Fig. 1C). The smaller change in the rate of DNA synthesis that occurs at termination (at about 2.75 genome equivalents) is masked by the biological and methodological variations just mentioned.

Thus, the DNA histograms of rapidly growing cultures are more complex than those of slowly growing cells because of the overlapping replication cycles. The information about the duration of C and D periods may be extracted by using either a computer simulation routine constructed for that purpose [6] or calculated from histograms of drug-treated cells [9] (see Sec. III. A).

During balanced growth in a defined medium, the rod-shaped *E. coli* cell doubles in mass and in length through the cell cycle, whereas the diameter does not change significantly. If the medium is changed so that it will support faster growth, the average diameter, length, and cell mass in the culture increase, as does the DNA content per cell. The cells in Figure 1F and H are drawn to correspond to cells with the same growth rate as those in Figure 1B and D, respectively. The light scattered at low-scattering angles, is a fair measure of cell size or mass [10]. The light-scatter histograms in Figures 1E and G are experimental ones from cultures with approximately corresponding growth rates, and are included to illustrate the size distributions obtained by flow cytometry. Interpretations of light-scatter histograms should be made with care, however, since this parameter also is dependent on shape, internal structure, absorption, and refractive index.

The cell mass at the time of initiation divided by the number of origins to be initiated is defined as the *initiation mass* [11]. The newborn cell in Figure 1F has a mass equal to the initiation mass because it contains a single chromosome that is being initiated (see Fig. 1B). The cell depicted at 15 min in Figure 1H contains four origins at which initiation occurs (see Fig. 1D). The initiation mass, therefore, is the cell mass at 15 min divided by 4.

III. FLOW CYTOMETRIC ANALYSIS OF THE *ESCHERICHIA COLI* CELL CYCLE

A. Determination of Initiation Age and Initiation Mass

Information about the cell cycle may be obtained by the use of drugs that inhibit initiation of DNA replication. Initiation, but not elongation, requires both transcription and de novo protein synthesis. These processes are inhibited by rifampin (rif) and chloramphenicol (cap), respectively. Addition of either drug to an exponentially growing culture leads to a gradual accumulation of cells with integral numbers of chromosomes. After "run-out" of replication (i.e., after all replication forks have reached their termini), a culture grown in rich medium contains cells with four or eight fully replicated chromosomes (Fig. 2B). If the cell division inhibitor cephalexin (cpx) is included together with the initiation inhibitor, division is also blocked, leading to a higher proportion of eight-chromosome cells (see Fig. 2C). The cells that ended up with four chromosomes

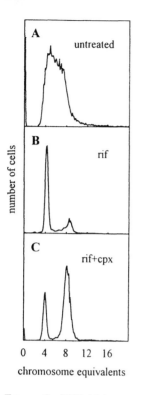

number of cells

0   4   8   12   16

chromosome equivalents

FIGURE 2 DNA histograms of (A) an exponentially growing *E. coli* K-12 culture, (B) treated for 4 h with rifampin alone, or (C) with both rifampin and cephalexin. The cells were stained with MI and EB.

were the youngest in the culture, and had not initiated DNA replication at the time of drug addition (see Fig. 1D), whereas the cells that ended up with eight chromosomes had initiated replication. Thus, the number of origins per cell in an exponential culture is found from such a rif/cpx histogram.

The DNA histogram of a rif/cpx-treated culture may also be used to determine the cell age and the cell mass at which initiation occurs [9]. The procedure is as follows: The fraction of cells that had not yet initiated replication at the time of drug addition is found from the number of cells in the four-chromosome peak (Fig. 3A). This information (fraction uninitiated) is combined with the cell age distribution (see Fig. 3B) to obtain the age, a_i (time in the cell cycle) at which cells initiate. This is done simply by setting the leftmost fraction in the age distribution (see shaded area, Fig. 3B) equal to the fraction uninitiated. The average mass at initiation, m_i, is found in a similar way, by combining the

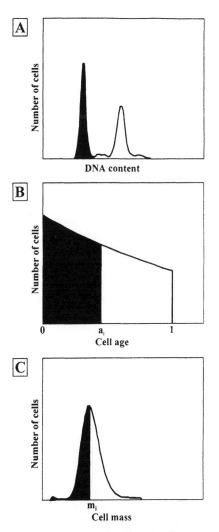

FIGURE 3 Determination of cell age and mass at initiation of DNA replication. (A) A DNA histogram of cells in a culture treated with rif and cpx. The cells with the lowest amount of DNA (shaded peak to the left) had not initiated replication at the time of drug addition. These cells were assumed to be both the youngest and the smallest in the population. The shaded portion in the theoretical age distribution (B), therefore, represents the same cells as the shaded fraction in panel A. The age at which the cells initiate replication is, thus, a_i. (C) In the cell size distribution of untreated cells, the same fraction can be found among the smallest cells, and used to find the average mass at initiation of chromosome replication (m_i) [9, 26].

information on the fraction of uninitiated cells (see Fig. 3A) with the size (light scattering) histogram of the untreated culture (see Fig. 3C). The fraction of smallest cells may be assumed to be the ones that had not initiated replication. Thus, by setting this fraction (see shaded area, Fig. 3C) equal to the fraction uninitiated, the average mass at the time of initiation, m_i, is found. (Dividing m_i by the number of origins at initiation then yields the parameter termed the initiation mass.

B. Timing of Initiation of Replication

Initiation of replication is precisely controlled and occurs at a certain time in the cell cycle. When two or more origins are initiated in the same cell, they are initiated simultaneously [12]. This was found by employing the method based on run-out of DNA replication, as described earlier (see Fig. 2). If initiation does not occur at exactly the same time at all origins within a cell, a fraction of the cells are caught with some origins, but not all four, initiated at the time of rif and cpx addition. This asynchrony leads to cells containing five, six, or seven fully replicated chromosomes after replication run-out, rather than eight. The longer the time interval between initiations, the more cells end up with numbers of fully

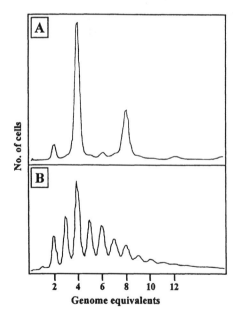

FIGURE 4 DNA histograms of (A) *E. coli* K-12 wild-type and (B) *dnaA46* mutant cells grown at 30°C to an OD_{600} of 0.2, treated for 4 h with rif and cpx, and then fixed and stained with MI and EB. (From Ref. 17.)

replicated chromosomes other than four or eight. In wild-type cells, initiation at all origins occurs almost simultaneously [12], and few cells end up with 3, 5, 6, or 7 chromosomes (see Fig. 2B,C and 4A). Certain mutations lead to aberrant initiation control [13–16]. Replication run-out histograms of, for instance, *dnaA* mutants, exhibit peaks at "odd" numbers of chromosomes (see Fig. 4B), indicating that the coordination of initiation of replication between the origins within each cell is lost [13]. In the *dnaA* mutants, however, it is unclear whether the signal to initiate replication is given at the right time in the cycle and the DnaA protein is just deficient in coordinating and initiating all origins, or whether it is the timing mechanism itself that is affected.

IV. CELL CYCLE ANALYSES OF OTHER MICROORGANISMS

Extensive analysis of bacterial cell cycle parameters has so far been concentrated on *E. coli*. Cell cycle analysis of other bacteria may be performed in a manner similar to that described in the foregoing. For species that tend to form chains, rather than to separate after cell division (e.g., *Bacillus subtilis*), it is important to sonicate the cell suspension to obtain a population of single cells.

The sensitivity of the microscope-based flow cytometer is useful also for obtaining high-resolution histograms of yeast cells. Figure 5 shows the DNA histogram of a *Schizosaccharomyces pombe* culture grown to stationary phase. Such high-quality histograms may aid the analysis of the many yeast cell cycle mutants that are available.

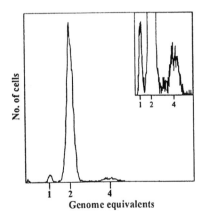

FIGURE 5 DNA histogram of stationary phase *Schizosaccharomyces pombe* cells, fixed and stained with MI and EB, as described. (Insert) Part of the same histogram at a higher magnification to illustrate instrument resolution.

V. EXPERIMENTAL PROCEDURES

A. Growth and Sampling

Cells should be grown for at least ten generations, to an optical density (OD) of no more than 0.2, measured at 450 nm in minimal medium and at 600 nm in rich medium [17]. In batch cultures of bacterial cells, the cell cycle parameters change continuously through the later part of exponential phase [5]. To achieve good reproducibility, therefore, it is recommended always to sample the culture at exactly the same OD. For bacteria, 1 ml of sample will suffice, whereas for yeast culture samples 10 ml is collected.

B. Fixation and Staining

For staining DNA, in most of our studies, we have used a combination of plicamycin (mithramycin; MI) and ethidium bromide (EB) [17,18]. Plicamycin is DNA-specific, but the fluorescence yield is low (about 0.05% [19]). Therefore, it is used in combination with EB to obtain a higher fluorescence yield. Ethidium bromide has negligible absorption at the excitation wavelength used (mainly the strong 436-nm line of the mercury arc lamp) and, therefore, is excited primarily by energy transfer from MI molecules bound nearby (< 5 nm away [19]). Because of the higher fluorescence yield of EB, energy transferred from MI to EB yields three to four times more fluorescence than can be obtained with MI alone. For measurements of bacteria, fluorescence from the RNA-bound EB is negligible at this excitation wavelength. However, for DNA measurements of yeast cells, the RNA must be removed by treatment with RNaseA (see following). We have also used the Hoechst 33258 and 4', 6-diamino-2-phenylindole (DAPI) dyes, with good result for DNA staining of bacteria and yeast. [For a review of DNA-staining procedures, see Ref. 20.]

Bacterial samples (1 ml) are harvested at an OD of 0.1–0.2, washed in cold TE (10 mM Tris, pH 8.0, 1 mM EDTA) by centrifugation for 3 min at 10,000 g in a standard refrigerated microcentrifuge and resuspended in 100 µl of cold TE. One milliliter of cold 77% ethanol is added to the suspension and mixed by vortexing. The cells may be stored at 4°C in the fixing solution for a few months, and at –20°C for much longer.

Yeast samples (10 ml) are also harvested at an OD of 0.1–0.2, chilled, centrifuged for 5 min at 4000 g, and resuspended in 1 ml cold TE; 9 ml of 77% ethanol is added during vortexing.

Before flow cytometry, fixed cells are washed once in staining buffer (10 mM Tris, pH 7.4, 10 mM MgCl$_2$). The cell pellet is then resuspended in ice-cold staining buffer. The cell density at this stage should be about 10^8/ml. Equal volumes of cell suspension and staining solution (180 µg/ml MI and 40 µg/ml EB in staining buffer) are mixed to ensure a reproducible dye concentration in the samples. The cells should be protected from light and kept on ice in staining solution for 30 min before measurement. Stained samples may be stored on ice for several hours before any deterioration (in terms of reduced quality of the measuring data, probably owing to DNA degradation) can be observed.

To obtain satisfactory DNA histograms of MI- and EB-stained yeast cells, treatment with RNaseA is necessary. Fixed cells are resuspended in 1 ml TE after centrifugation at 4000 g for 5 min, and incubated at 37°C for 60 min in a solution of 100 µg/ml RNaseA and 50 mM EDTA in 10 mM Tris–Cl, pH 8.0. The tube is flicked every 15 min to keep the cells in suspension. The cells are then centrifuged again and resuspended in staining buffer and stained as described earlier. Yeast cells often stick together, yielding incorrect DNA and size histograms. Therefore, it is important to sonicate the stained cells, so that a suspension of single cells is obtained. The extent of sonication necessary should be checked with microscopy. Excessive sonication must be avoided because it leads to cell breakage.

C. Instrumentation

The type of flow cytometer best suited for measuring small microbial cells exploits the principle of the epifluorescence microscope [21]. An outline of this instrument (Argus 100, Skatron AS, Lier, Norway) is given in Figure 6. A newer version of the instrument is currently available from BioRad (Bryte HS, BioRad Spd., Milano, Italy). The excitation light source is a 100-W, high-pressure mercury lamp, the arc of which is imaged on the "excitation slit" which, in turn, is imaged in the object plane of a microscope lens. The light from the arc lamp is passed through a band-pass interference filter transmitting the wavelength used to excite the staining dye. This light is reflected by a dichroic mirror into a microscope lens that concentrates the light in the focus through which the sample flow is passing. Fluorescence from the cells is collected by the same lens and passes through the dichroic mirror (because of its longer wavelength, compared with the excitation light) and onto the "measuring slit," which is situated in the image plane of the microscope lens. Thus, the fluorescence light is collected in the direction opposite that of the excitation. Since this is the direction where the intensity of the stray light from the flow chamber, as well as from the lens itself, is lowest, this optical configuration produces a low level of background, which is essential to achieve sufficient sensitivity. The excitation and measuring slits eliminate stray light and, thereby, further reduce the amount of background light that reaches the detectors. To concentrate a maximum intensity of excitation light in the focus and to collect as much fluorescence as possible, the lens is a microscope objective with oil immersion and a numerical aperture (NA = 1.3) close to the theoretical maximum.

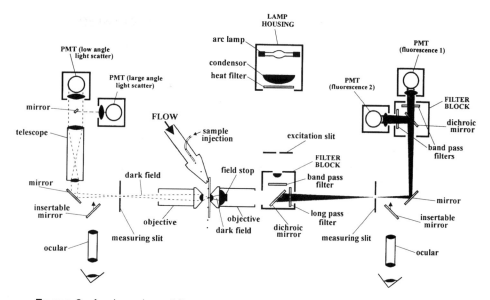

FIGURE 6 Arc lamp-based flow cytometer. Fluorescence is detected in epimode, which means that the same microscope objective is employed to concentrate the excitation light on the sample flow and to collect fluorescence. The excitation and fluorescence wavelengths are selected by the combination of excitation filter, dichroic mirror, and long-pass filter, which are situated in an interchangeable filter block. A second filter block in front of the photomultiplier tubes permits separate collection of two different wavelengths. Light scattering is detected in a dark-field configuration. A secondary microscope collects scattered light within the dark field produced by a field stop in the primary microscope objective. The dark field is reproduced behind the measuring slit by means of a telescope to facilitate separate detection of light scattered to small and large angles.

Detection of scattered light is carried out in a dark-field configuration [22]. The microscope lens contains a central field stop that produces a conical shadow pointed in the focus and extending on the opposite side of the object plane. This shadow contains only that light originating in the focus; that is, fluorescence and scattered light. A lens having its object plane in common with the first lens and its aperture within this shadow, collects only fluorescence and scattered light. The fluorescence, which is generally much weaker than the scattered light, can be eliminated by an appropriate filter, so that the secondary lens forms a scattered-light image of the cell flow on a measuring slit, which eliminates stray light and reduces background by the same principle as in the fluorescence detection. In the scattered-light field behind the slit, it is possible to divert light scattered to low and high angles to separate detectors [23,24]. Light scattered to low angles is a measure primarily of cell mass [10], whereas scattering to higher angles depends more on cell shape and structure.

Fluorescence and light-scattering detection is performed by photomultiplier detectors that transform the light pulses to equivalent electrical signals that are digitized and stored as histograms in a computer. The DNA and size distributions may be displayed separately, or combined into dual-parameter histograms.

The flow chamber is of the "jet on open surface" type. The jet from a nozzle with hydrodynamic focusing impinges on the open surface of a cover slip to produce a flat, laminar flow, with the cells confined to a narrow sector along the middle of the flow. This type of flow chamber exhibits fewer interfaces than closed-flow chambers and, therefore, gives rise to less stray light and a correspondingly lower background [24]. The flow velocity is of the order of 10 m/s.

VI. STANDARDS AND CONTROLS

Monodisperse fluorescent beads are run when the instrument is set up, to check the nozzle and water jet, and to optimize the instrument for sensitivity and resolution. The coefficient of variation (CV) of the fluorescence peak should be less than 2%, preferably as low as 1.2%. A standard cell sample is run to monitor staining and to calibrate the DNA axis. The sample is usually rifampin-treated bacteria (see Fig. 2), yielding sharp peaks at integral numbers of chromosomes. The standard is run at regular intervals between samples. If calibration of the DNA axis is especially critical, for instance, if absolute DNA content per cell is required, rather than relative values, it is recommended that more than one type of calibration sample is run (e.g., chloramphenicol-treated cells or a *dnaA* (Ts) mutant incubated at nonpermissive temperature, in addition to rifampin-treated cells). All the calibration samples should give a similar amount of fluorescence per chromosome. For continuous monitoring of instrument stability, inclusion of beads in all samples is recommended.

ACKNOWLEDGMENTS

We are grateful to Mette W. Jernæs for comments on the manuscript. This work was supported by grants from The Research Council of Norway (K. S.), The Norwegian Cancer Society (E. B.) and The Swedish Natural Science Research Council (R. B.).

REFERENCES

1. E. Boye, and A. Lobner-Olesen, Flow cytometry: illuminating microbiology, *New Biol.* 2:119–125 (1990).

2. P. Fouchet, C. Jayat Y. Héchard, M.-H. Ratinaud, and G. Frelat, Recent advances of flow cytometry in fundamental and applied microbiology, *Biol. Cell* 78:95–109 (1993).

3. R. M. Ferry, L. E. Farr, Jr., and M. G. Hartman, The preparation and measurement of the concentration of dilute bacterial aerosols. *Chem. Rev.* 44:389–395 (1949).

4. J. E. Bailey, J. Fazel-Madjlessi, D. N. McQuitty, L. Y. Lee, J. C. Allred, and J. A. Oro, Characterization of bacterial growth by flow microfluorometry, *Science* 198:1175–1176 (1977).

5. K. Skarstad, H. B. Steen, and E. Boye, Cell cycle parameters of slowly growing *Escherichia coli* B/r studied by flow cytometry, *J. Bacteriol.* 154:656–662 (1983).

6. K. Skarstad, H. B. Steen, and E. Boye, *Escherichia coli* DNA distributions measured by flow cytometry and compared with theoretical computer simulations. *J. Bacteriol.* 163:661–668 (1985).

7. S. Cooper and C. E. Helmstetter, Chromosome replication and the division cycle of *Escherichia coli* B/r, *J. Mol. Biol.* 31:519–540 (1968).

8. R. Allman, T. Schjerven, and E. Boye, Cell cycle parameters of *Escherichia coli* K-12, *J. Bacteriol.* 173:7970–7974 (1991).

9. S. Wold, K. Skarstad, H. B. Steen, T. Stokke, and E. Boye, The initiation mass for DNA replication in *Escherichia coli* K-12 is dependent on growth rate, *EMBO J.* 13:2097–2102 (1994).

10. E. Boye, H. B. Steen, and K. Skarstad, Flow cytometry of bacteria: a promising tool in experimental and clinical microbiology, *J. Gen. Microbiol.* 129:973–980 (1983).

11. W. D. Donachie, Relationship between cell size and time of initiation of DNA replication, *Nature* 219:1077–1079 (1968).

12. K. Skarstad, E. Boye, and H. B. Steen, Timing of initiation of chromosome replication in individual *Escherichia coli* cells, *EMBO J.* 5:1711–1717 (1986).

13. K. Skarstad, K. von Meyenburg, F. G. Hansen, and E. Boye, Coordination of chromosome replication initiation in *Escherichia coli*: effects of different *dnaA* alleles, *J. Bacteriol.* 170:852–858 (1988).

14. E. Boye, A. Lobner-Olesen, and K. Skarstad, Timing of chromosomal replication in *Escherichia coli*, *Biochim. Biophys. Acta* 951:359–364 (1988).

15. E. Boye, and A. Lobner-Olesen, The role of *dam* methyltransferase in the control of DNA replication in *E. coli*, *Cell* 62:981–989 (1990).

16. M. Lu, J. L. Campbell, E. Boye, and N. Kleckner, SeqA: a negative modulator of replication initiation in *Escherichia coli*, *Cell* 77:413–426 (1994).

17. K. Skarstad, R. Bernander, and E. Boye, Analysis of DNA replication in vivo by flow cytometry, *Methods in Enzymology, DNA Replication* (J. Campbell, ed), Academic Press, New York (in press)

18. H. B. Steen, M. W. Jernaes, K. Skarstad, and E. Boye, Staining and measurement of DNA in bacteria, *Methods in Cell Biology* (H. A. Crissman and Z. Darzynkiewitz, eds), Academic Press, New York (in press).

19. R. G. Langlois, and R. H. Jensen, *J. Histochem. Cytochem.* 27:72–78 (1979).

20. H. M. Shapiro, *Practical Flow Cytometry*, 2nd ed., Alan R. Liss, New York, 1988.

21. H. B. Steen, and T. Lindmo, Flow cytometry: a high resolution instrument for everyone, *Science* 204:403–404 (1979).

22. H. B. Steen, and T. Lindmo, Differential light-scattering detection in an arc-lamp-based epi-illumination flow cytometer, *Cytometry* 6:281–285 (1985).

23. H. B. Steen, Simultanous separate detection of low and large angle light scattering in an arc lamp-based flow cytometer, *Cytometery* 7:445–449 (1986).

24. H. B. Steen, Characteristics of flow cytometers, *Flow Cytometry and Sorting*, 2nd ed. (M. R. Melamed, T. Lindmo, and M. L. Mendelsohn, eds.), Wiley-Liss, New York, 1990 pp. 11–25.

25. W. D. Donachie and A. C. Robinson, Cell division: parameter values and the process, Escherichia coli *and* Salmonella typhimurium *Cellular and Molecular Biology* (F. C. Neidhardt, J. L. Ingraham, B. Magasanik, K. B. Low, M. Schaechter, and H. E. Umbarger, eds.), American Society for Microbiology, Washington, DC, 1987 pp. 1578–1593.

26. E. Boye, and H. B. Steen, The physical and biological basis for flow cytometry of Escherichia coli, *Flow Cytometry in Microbiology* (D. Lloyd, ed.), Springer Verlag, London, 1993 pp. 11–25.

15

Bacterial Characterization by Flow Cytometry

GERHARD NEBE-VON CARON AND R. ANDREW BADLEY
Unilever Research, Sharnbrook, Bedford, England

I. INTRODUCTION

The rise of biotechnology in its many guises over the last decade or so has acted as one of a number of spurs to the development of new analytical tools for the characterization of cells of all types. To be able to follow such features as growth, death, secretion, cell division, metabolism, and surface phenomena, just as some examples, enables both understanding and control of cell behavior. Flow cytometry has been in the forefront of this drive. The first attempts at bacterial detection were made as early as the late 1940s, driven by the need to identify bacterial aerosols in warfare [1,2]. The next period of more intensive flow cytometric study of bacteria was in the late 1970s [3–6]. Hutter and Eipel were the first to undertake a complex study on viability, total protein, and cell cycles of bacteria, yeasts, and molds, and the autofluorescence of algae. They had already employed the power of multiparameter measurements possible with flow cytometry, a feature neglected in many of the more recent studies. At the same time Steen used a modified microscope that he developed into a flow cytometer more geared for microbial applications. He did fundamental work in bacterial replication and, subsequently, in drug susceptibility [7]. Aspects of these and more recent developments have been reviewed [e.g., 7–12].

The ability of flow cytometers to make multiparameter measurements sets them apart from most other cell analytical techniques. Thus, one of the key features we should expect to be able to analyze for bacterial populations is the nature of any heterogeneity. This is possible, as we shall see later, but progress has certainly been slower than for the generally larger eukaryotic cells that have been studied. A principal reason for the slow progress lies in the size of typical bacteria. They are approximately ten times smaller in terms of their diameter than typical white blood cells, for which many modern flow cytometers have been designed. This means that the corresponding surface area is 100 times less, and the volume is 1000 times less. The signals from surface-bound or volume-bound probes will also be decreased in the same proportions. Consequently, sensitivity has been, and still is, *the* key issue for many bacterial flow cytometric studies, especially for standard commercial instrumentation.

Thus, two important themes will permeate the discussion in this review of bacterial characterization by flow cytometry; multiparameter measurements of bacteria are used to reveal heterogeneity in particle mixtures containing bacteria, and bacterial size often leads to issues of sensitivity.

II. THE PRESENCE, ABSENCE, AND COUNTING OF BACTERIA

The classic culture technique in microbiology is a very sensitive and highly amplified (10^{9-12}) single-cell detection system. The strength of the method is also its limitation, for it is dependent on post sampling growth. However, it provides a piece of useful information, as it tells us about reproductive viability under laboratory conditions. Difficulties arise with highly adapted, stressed, or injured cells, for which recovery depends on finding the right growth conditions, or for which the time required for recovery is critical. Also, vital, but nonculturable, cells represent a major problem in microbiology. Finally, culture technique gives only a limited understanding of cell physiology or mode of action for stress or injury. Therefore, it is best to aim to measure "on-line" and, if possible, still on a single-cell level.

A crude way to detect bacteria is based on the measurement of light-scatter signals only. It allows detection of particles within a certain size range. Applied to pure cultures, in an otherwise particle-free environment, this is sufficient for bacterial counting and, thus, detection of reproductive growth. Because the scatter is a function of particle size and internal structure, it can sometimes be used to differentiate bacteria from other materials, or from each other, as well as to detect aggregation. Thus, it is not surprising that changes in light scatter during batch culture allow flow cytometry to give growth kinetic determinations superior to those given by optical density measurements. [13; R. Allman; H.B. Steen, personal communications].

Counting statistics are also in favor of flow cytometry. By counting more than 5000 events, it is possible to achieve counting errors below 1%, allowing detection of subtle changes. This has been exploited in hematology counters, originally measuring conductivity (Coulter counter), but now also light scatter. In most applications, flow cytometry is used only to analyze percentages of subpopulations, whereas for microbiology, absolute counts are still taken by microscopy [14–16]. Until recently, even in clinical immunology, absolute counts from hematological analyzers had to be combined with the relative cell frequencies. However, acquisition of absolute total counts of bacteria is essential when investigating lag phases, dormant or nonculturable cells, especially when physiological measurements are taken [15]. To turn detected events directly into absolute numbers per volume [17–20], there are three possible approaches:

1. *Fixed volume counting or volume integration:* This method is used in most hematological analyzers. The volume measurement is achieved by two electrodes acting as level sensors in a known geometric setup or by loading a cavity in a sheer valve or a loop made of tubing, as done in high-performance liquid chromatography (HPLC) instruments. In both procedures all events within the volume are measured. Problems can occur with leakages, or if too few or too many particles are present in that volume.

2. *Time integration:* This approach is based on the assumption of a constant volume flow over time. It is best achieved by (syringe) pumps delivering the sample. Such systems are implemented in the Ortho Cytron Absolute and the former Skatron Argus flow cytometer, now sold by Biorad. Problems can potentially occur with leakages or blockages in the system, pump speed instabilities, and sample carry-over.

3. *Spiking with reference particles (ratiometric method):* Mixing known volumes of solutions of reference particles and unknown sample allows calculation of absolute counts from the measured particle ratio. Difficulties lie in aggregation or coincidence of beads and bacteria or elimination of particles by phagocytosis. The benefit of the method is that it can be used with all types of cytometers, even for crude counts by microscopy. The major advantage of the ratiometric method is that it compensates for flow rate changes and system dead times. It also acts as an on-line quality control (QC) method and allows a certain freedom of sample manipulation in terms of washing and dilution steps [21]. Ortho Diagnostics, Becton Dickinson, and Coulter Corporation are currently starting to release counted bead standards for use in clinical immunology.

Spiking with reference particles has been examined as a means of counting bacteria routinely by flow cytometry. To facilitate easier detection of small particles, including bacteria, a Coulter Epics Elite flow cytometer was modified so that forward-scatter light was rerouted to a photomultiplier by a short length of fiber optic cable. Most bacteria can be detected by the built in detector, but such a modification could be considered for difficult cases. An example for

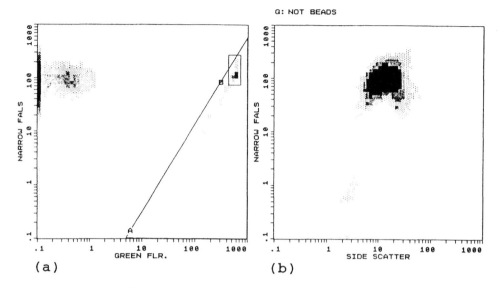

FIGURE 1 Counting of bacteria by flow cytometry using reference beads: (a) The beads are identified in a dot plot of narrow forward-angle light scatter (FALS, *y*-axis) versus green fluorescence (*x*-axis). Region B includes the 660-nm beads, region A also the smaller beads included in the preparation. (b) Narrow FALS (*y*-axis) versus right-angle scatter (RALS or side scatter) confirms the clustering of the bacteria in light scatter. All green fluorescent beads from region A have been excluded by gating.

bacterial counting using spiked reference beads is given in Figure 1. The cluster of the standard (660-nm diameter) is identified in region B on the basis of size and fluorescence. All beads including the smaller fraction (region A) are excluded (gated out) in the light-scatter dot plot.

Clearly, for samples containing interfering particles, the use of light scatter alone is insufficient for detecting or counting bacteria. In these cases additional parameters need to be measured to identify the bacteria of interest. Most commonly, the presence of nucleic acid is used as a target for specific stains but, in principle, any of the many bacterial stains (see rest of chapter) can be used. The absence of nucleic acid in interfering particles has proved very effective in practice.

Figure 2 gives an illustration of particle interference when counting bacteria and, thus, of the limitations of light-scatter detection. In the light-scatter dot plot (see Fig. 2a), the reference beads have been excluded as before. In Figure 2b, bacteria are positive to the nucleic acid stain ethidium bromide (EB) in region C (right box), which discriminates them from the unstained interfering

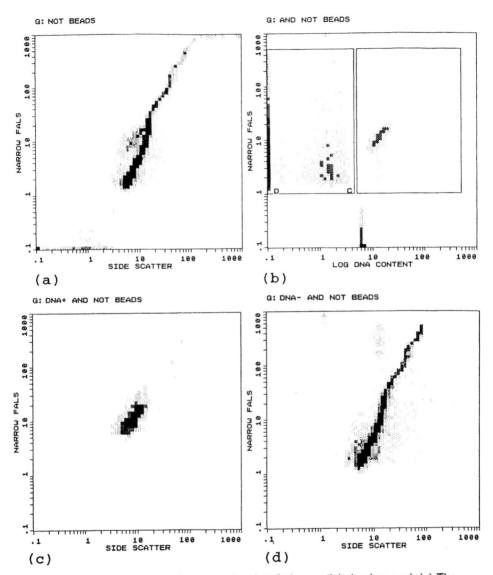

FIGURE 2 Bacterial discrimination against interfering particle background: (a) The light-scatter dot plot after exclusion of the reference beads. The bacterial cluster is masked by interfering particles. (b) The separation of events in narrow FALS (*y*-axis) versus red DNA fluorescence (*x*-axis) of ethidium bromide-stained cells (region C, right box). (c,d) The light-scatter dot blots gated on regions C and D, which demonstrates the discontinuous distribution of the DNA-positive events typical for bacteria and the tail-like distribution of the interfering debris and micelles.

particles (region D, left box). The light-scatter dot plots gated on these regions are shown (Fig. 2c,d), to show they cluster separately.

To check the reproducibility of ratiometric counting and to compare it with time integration counting, 15 aliquots of a bead and bacteria mixture were measured. Six such mixtures were compared, containing different additives. Two of the detergents (Decon 90 and Tego Betaine) were discarded, because one caused a precipitate to form, and the other killed the cells. Skimmed milk powder as a blocking agent caused severe problems with particle interference. The results of these counts are given in Table 1. The counts per minute of bacteria (column 2) and beads (column 3) show significantly higher variation compared with the ratiometric counts. The theoretical CV_{RATIO} values calculated as $[CV_{RATIO} = $ square root of $(CV_A^2 + CV_B^2)]$ are based on the mean of the counting errors (= square root of count) of beads and bacteria. The variation between the counting errors of beads and bacteria in the 15 replicas was 10% or less.

TABLE 1 Comparison of Bacterial Counts Obtained by Time Integration and the Reference Particle Ratio Method

	Counts/min bacteria[a]	Counts/min beads[a]	Concentration ratio of bacteria[a]
PBS-Tween			
Mean	5562	1005	5.53
SD(n–1)	1198	207	0.21
%CV (n–1)	21.5	20.6	3.8
Theoretical CV_{RATIO}		3.5	
PBS-Pluronic			
Mean	5816	1095	5.32
SD(n–1)	513	108	0.20
%CV(n–1)	8.8	9.9	3.80
Theoretical CV_{RATIO}		3.30	
PBS–BSA			
Mean	5429	1065	5.1
SD(n–1)	788	144	0.25
%CV(n–1)	14.5	13.5	4.8
Theoretical CV_{RATIO}		3.4	
Individual sample preparations in PBS–Pluronic			
Mean	6063	1191	5.1
SD(n–1)	700	136	0.23
%CV(n–1)	11.5	11.4	4.5
Theoretical CV_{RATIO}			3.2

[a]Aliquots taken from one 10-ml spiked sample buffer (~10^9/ml)

   Thus, particle counting by ratiometric bead count has proved to generate far lower coefficients of variation (CV) than time integration, but the latter's poorer performance in our flow cytometer may have been caused by small losses of air pressure at the tube seal and variations in sample run time before the acquisition started. In the presence of detergents the observed CVs are less than 1% above the predicted values, based on the counting error of the actual counts.

   To investigate the linearity of the system, a dilution curve of *Streptococcus sanguis* was counted. The correlation between sample dilution and ratiometric counts by flow cytometry is shown in Figure 3. The coefficient of regression was calculated as $r^2-0.999$. The lowest concentration counted was limited by the simultaneous acquisition of non-bacterial counts filling up the computer memory.

   Even when measuring pure cultures, cell aggregation is an important problem for counting accuracy, and more so with samples such as yoghurt or bacterial plaque, for which aggregation is easily verified by optical microscopy. Aggregation in chains or clusters can be species-specific, or can depend on growth conditions. To use the power of single-cell measurements, the presence of single cells is essential and is usually achieved by various mechanical or chemical disruptions. Counting achieved by such treatments can cause discrepancies, compared with conventional counting methods, especially with samples

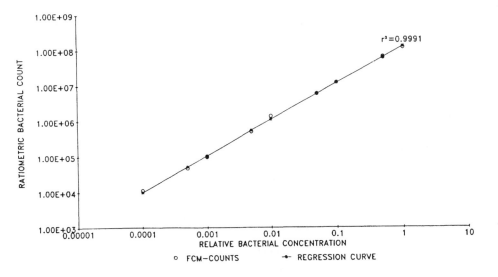

FIGURE 3 Linearity of counts obtained by flow cytometry: Bacteria were diluted in different ratios down to 10^{-4} and counted by flow cytometry to show the correlation between the dilution and the flow cytometric counts. The coefficient for the linear regression was $r^2 = 0.9991$.

not prepared in such a way. This can lead to questions about what should be the gold standard.

Occasionally the cells have to be released from other sample material by some form of homogenizing treatment. Alkaline hydrolysis, which is used to remove RNA of fixed cells, facilitates cell separation in sewage [A. Völsch, personal communication], and washing in milk-clearing solution (a reagent used to prepare milk samples for automated detection of contaminating cells) can remove particles interfering in egg samples [A. Pinder, personal communication].

To disturb metabolic function as little as possible, mechanical separation should be superior and more general than chemical methods. Because of the limited disaggregation caused by standard needles and the blocking problems encountered with squashed needles during earlier work on sewage, the capabilities of sonication were investigated.

TABLE 2 Effect of Sonication on Cell Integrity and Separation/Recovery of Aggregated Rods and Cocci

Starter culture	Treatment		Count (10^7/ml)	Membrane integrity (%)
Rods				
A8 overgrown	4 min	2 μm	1.9	54
	2	2	2.0	53
	1	2	1.5	41
	20 s	Bath	0.9	38
	Untreated		0.7	31
A8 fresh	4 min	2 μm	18	96
	2	2	21	97
	1	2	18	97
	20 s	Bath	10	98
	Untreated		Too aggregated	—
Cocci				
SB14 overgrown	4 min	2 μm	0.9	16
	2	2	0.8	10
	1	2	0.8	5
	20 s	Bath	0.5	8
	Untreated		0.3	7
SB14 fresh	4 min	2 μm	23	67
	2	2	21	68
	1	2	19	68
	20 s	Bath	10	66
	Untreated		Too aggregated	—

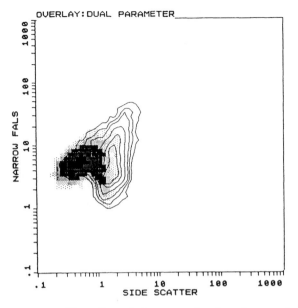

FIGURE 4 The effect of disaggregation by ultrasound on light-scatter signals: A contour plot of narrow FALS (*y*-axis) versus side scatter (*x*-axis) of *Peptostreptococcus anaerobis* grown under argon is overlaid with a dot plot of the same sample after sonication for 10 s in the hot spot of a water bath. The remaining short chains after sonication still cause a slight tail toward higher side scatter but lower narrow FALS.

The effect of sonication on different bacterial shapes was tested on bacterial isolates SB14 (cocci) and A8 (rod). Both species were grown overnight in a tryptone yeast extract, to maximize the aggregation problem. A 1-ml sample was sonicated in a 5-ml Sterilin polystyrene container, using an exponential microprobe, with a 3-mm-tip diameter, set to 2-μm amplitude, at 23 kHz for different lengths of time, or it was held in the hot spot of a sonicating water bath. The results are shown in Table 2. The overnight cultures inoculated with overgrown colonies of both species were still in exponential growth, whereas cultures from fresh colonies were already aggregated.

The results obtained from separation by sonication show the suitability of this method for the two representative species chosen. Membrane integrity measurements using propidium iodide (PI, see Sec. IV) did not indicate a higher sensitivity of the rods to sonication, but this might be due to the relatively short length achieved under these growth conditions. Rods and cocci show maximum recovery at 2-min sonication at 2-μm amplitude. The apparent increase in intact cells with prolonged sonication is mainly due to separation of cells (one stained cell attached to four unstained ones makes the whole aggregate look stained), but might indicate destruction of dead cells as well. Real filamentous bacteria seem to be more sensitive to sonication, as observed by microscopy. Preparation of such cells, therefore, may require a more gentle approach, possibly sheering in a needle. Experiments on the obligate anaerobe *Peptostreptococcus anaerobis* indicate that it also survives the 2-min sonication, with no loss in viability.

With errors of flow cytometric counts less than 3%, cell separation becomes the accuracy-limiting step for counting bacteria. With the possible destruction of some cell types, separation methods have to be optimized for the organisms of interest. Cell separation also allows the assessment of the degree of aggregation or the energy required to separate the cells as a form of culture characterization. It also has a strong influence on the signals derived from the cells, as demonstrated in the overlay light-scatter dot plot of *P. anaerobis* (before sonication) and light-scatter dot plot (after sonication; Fig. 4).

III. MORPHOLOGY AND LIGHT SCATTER

Traditionally, for flow cytometrists, there is a clear link between size and so-called forward-angle light scatter (FALS), on the one hand, and between internal structure–shape (morphology) and right-angle light scatter (RALS or side scatter), on the other. This has arisen from the major use of flow cytometry in the study of mammalian cells, especially blood cells. They are large enough, compared with the wavelength of the incident light, and spherical enough to behave in a relatively ideal manner, although, even here (to the practiced eye), subtle differences in both types of light scatter can be used to differentiate similar-looking cell types.

For bacteria, the assumptions are definitely not valid. This is well illustrated by comparing the light-scatter dot plots from *Bacillus cereus* and those of its spores (Fig. 5). The spores clustering with the high forward-scatter signal in region A are, in real terms, much smaller than the outgrowing cells (region B). This is partly due to the angular dependency of forward-scattered light and particle size, and partly to the higher refractive index difference between the spore and its aqueous environment.

There are two other factors that can distort the expected signals. First, a lack of orientation of nonspherical objects, such as long rods or chains of cocci. An example of the latter is given in Figure 4, with aggregated *P.* anaerobis in contour display and sonicated cells overlaid as a dot plot. The remaining short chains, after the brief sonication, still cause a tail in the plot to the lower right of the cluster. Second, the varying angles and ranges of angles that different cytometers use to collect forward and side scatter can make significant alterations to the observed results. This is demonstrated using latex beads, with a size range from 375 to 2230 nm (Fig. 6). Apart from the lower sensitivity of the wide

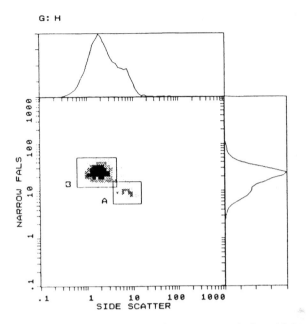

FIGURE 5 Forward-angle light scatter and size: Light-scatter dot plot and its projection for *Bacillus cereus* spores (region B) and outgrowing cells (region A). While under the microscope the spores appear smaller than the outgrowing cells, their narrow FALS signal is stronger, reflecting the influence of refractive index on FALS.

forward-angle light-scatter detector diode (see Fig. 6b), the relative signal of the 2230-nm beads (region L) is lower than for the 1160-nm–diameter beads (region K) for narrow FALS, but equal in wide FALS. Only the right-angle light-scatter signal shows constant increase with size (see Fig. 6a) and has been calculated to correlate with diameter to the power of 2.41 ($[diameter]^{2.41}$) with a correlation coefficient (r^2) of 0.9935 for these latex beads.

Despite these limitations, characterization of pure cultures by light-scatter measurements has been described by several authors [17,22–25]. Bronk [23] and Wyatt [22] showed the angular dependency of light-scatter signals and the differences between species and growth stages. Correlation of forward scatter has been achieved with Coulter counter sizing [17,26] and with electron microscopy and image analysis [26,27]. Allmann [28] has used light scatter in connection with DNA measurements and neural network data analysis to attempt differentiation of mixed populations. Gas vacuoles [29] and other cytoplasmic inclusions [30,31] can give changes in light-scatter patterns and, thus, lead to differentiation of bacteria otherwise impossible to achieve. The complexity of

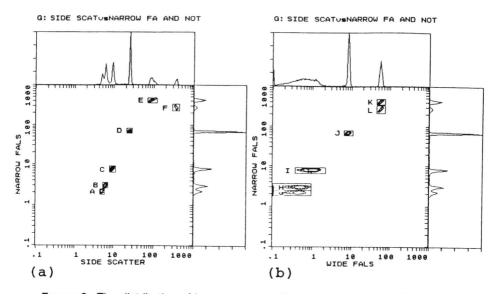

FIGURE 6 The distribution of latex beads at different scatter angles: The contour plots show the distribution of beads of increasing size from 375 to 2230 nm [(a) regions A–F and (b) G–L]. Although the ranking in side scatter correlates with diameter$^{2.41}$ ($r^2 = 0.994$), the 2230-nm beads (region F) give less signal than the 1160-nm beads (region E) in the narrow forward-angle light scatter (FALS) (a). When comparing narrow and wide FALS in (b), the relative position between the 1160-nm beads (region K) and the 2230-nm beads (region L) is significantly different. The same is true for the relative positions of the 660-nm beads and bacteria in the two forward-scatter detectors (data not shown).

light-scatter distributions is illustrated on an exponential culture of *Listeria monocytogenes* in Figure 7. The reproducible pattern in the distribution indicates the possibility of light–scatter-based differentiation within a culture, but also limits its application for the differentiation of mixed cultures.

IV. MEMBRANE INTEGRITY

The principal use of measurements of membrane integrity, or the lack of holes in the outer cell bilayer membrane, is to gain an indication of cell viability (see later). Eukaryotic and prokaryotic cell membranes are designed to prevent the uncontrolled passage of most molecules into the cell, and such penetration, therefore is, an indicator of cell damage, at the least, and cell death, at the extreme.

G:

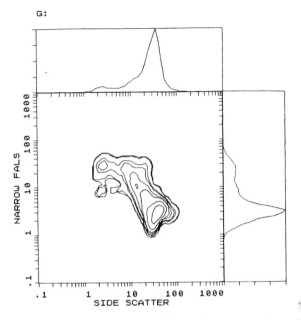

FIGURE 7 Light-scatter variability within a pure culture: The distribution in narrow FALS versus side scatter and their projections is shown for an exponentially growing culture of *Listeria monocytogenes*. A similar picture can be obtained with *Lactobacillus plantarum* or *E. coli*. The distribution pattern does not change under sonication, as there are few aggregates in exponential growth.

Many intracellular dyes used in flow cytometry depend on circumventing the cellular barrier by one means or another. Only a very few, mostly lipophilic, dyes can readily cross an intact membrane. Dyes that are normally impermeable to the cell membrane and have an intracellular binding site, therefore, can be used to detect membrane integrity by dye retention measurements or by dye exclusion measurements.

The loading of impermeable dyes for retention measurements can be achieved under nonphysiological conditions (acid shock, high solvent concentration, or electroporation) or by using membrane-permeant enzyme substrates that become impermeable to the membrane after the internal target enzyme has acted on them. Even when the dye loading has been achieved successfully, problems can still arise because the cells are still alive. The fluorochromes can be quenched (e.g. by changes in pH or by becoming too concentrated). They can even be removed actively from the cell by membrane-bound pumps, a factor that has to be kept in mind for all intracellular stains.

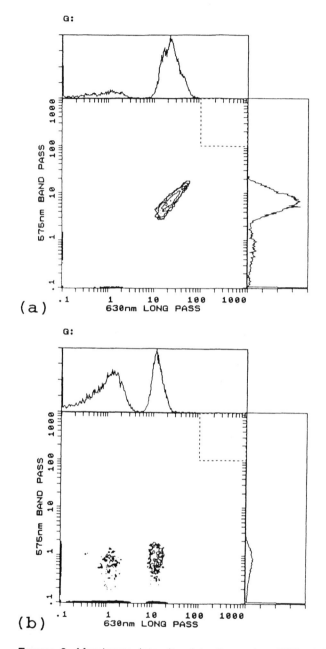

FIGURE 8 Membrane integrity detection using DNA stains: Discrimination of exponentially growing and heat-fixed *Listeria innocua* by the DNA stains (a) ethidium bromide (EB) and (b) propidium iodide (PI). Although PI stains only the

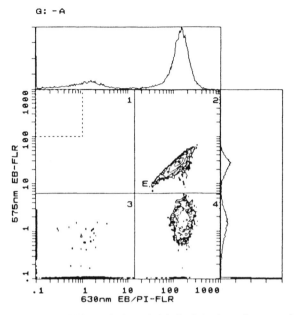

FIGURE 9 Differentiation of debris, intact, and permeabilized bacteria by simultaneous two-color staining with propidium iodide and ethidium bromide: An exponentially growing culture of *E. coli* strain D664b and a heat-fixed sample of the same culture were mixed and stained with ethidium bromide and propidium iodide (10 μg/ml each). Quadrant 3 contains debris particles that do not pick up any DNA stain. The cells stained with only ethidium bromide give rise to red and orange emission (see also Fig. 8b) and fall in quadrant 2. The heat-fixed cells that contain both fluorochromes have lost the orange component of their emission because the propidium iodide quenches the emission of the ethidium bromide. They appear in quadrant 4 of the picture.

Dye exclusion measurements are mostly based on nucleic acid stains that are impermeable to the intact membrane. Once they can freely enter the cell, the interference of pumps is unlikely. The most commonly used dye exclusion markers are ethidium bromide (EB) and propidium iodide (PI). Even for mammalian cells, uptake of EB has to be treated with caution [10, p. 168]. In some bacteria, active concentration of EB has been observed [32]. The different behavior of these two dyes is demonstrated on a mixture of heat-fixed and viable bacteria (Fig. 8). The fact that EB stains intact bacteria, unless pumped out, but

---

(**FIGURE 8** continued) heat-fixed population red (630 nm, *x*-axis), ethidium bromide stains all cells and has a slightly more orange emission (575 nm, *y*-axis). The unstained events (ethidium bromide-negative) can be identified as debris, as they result in a light scatter image similar to that in Figure 2d.

PI does not, may be used for positive discrimination of a selective (intact) membrane. When using both stains simultaneously (Fig. 9), the DNA of permeabilized cells is stained with both fluorochromes. This leads to energy transfer from the orange-emitting EB to the PI that gives only a far-red fluorescence [10, p. 168]. The same effect occurs when using ultraviolet-excited blue DNA stains, such as Hoechst, with PI quenching the blue fluorescence. The SYTO dyes from Molecular Probes might allow even better dye combinations for such measurements and could even escape removal by pumps if their affinity to DNA is high enough. The degree of permeabilization can be tested by further probing for the molecular size that can pass the membrane. This can be determined using fluorescent probes coupled to proteins of known molecular weight [33].

V. METABOLIC ACTIVITIES

In common with all other living cells, bacteria perform numerous chemical reactions, most of which use enzymes and require energy to perform. This dependence on energy is really a common theme, which the flow cytometrist can exploit in two ways: the characterization of particular cellular reactions within the cell for their own sake, or as an indicator of cell viability (see Sec. IX). The means by which such reactions are studied largely depends on our ability to design fluorescent dyes, the optical properties of which are altered, either by chemical changes, or by a change in concentration or environment. An examination of the literature for eukaryotic cells indicates a very wide and diverse range of metabolic activities that have been studied [e.g., 10, p. 181], but there are also significant gaps. For bacteria, fewer activities have been looked at because of associated technical difficulties; because of smaller signals from smaller cells, because some eukaryotic reactions do not occur in bacteria, or because of difficulties with staining techniques arising from different cell wall or membrane compositions. We have picked out three topics that are either already established areas of study, or that are likely to be in the near future: membrane potential, intracellular pH, and enzyme activities.

A. Membrane Potential

Membrane potential is probably one of the most frequently abused measurements in flow cytometry. It originates from the selective permeability and active transport through a membrane. The Nernst equation describes the relation between the distribution of a charged molecule across the membrane and its membrane potential.

$$E_i - E_o = (\ln a_o - \ln a_i)\frac{RT}{zF}$$

$$= \ln\left(\frac{a_0}{a_i}\right)\left(\frac{RT}{zF}\right)$$

where

E = electropotential

i = inside; o = outside the cell

a = active (diffusable or nonbound) concentration

R = Reynolds constant

T = absolute temperature

z = charge

F = Faraday constant

The most commonly used probes for membrane potential estimations are charged lipophilic dyes, such as carbocyanines (DiOC$_5$), rhodamine 123, or oxonols. Probes that can readily cross the membrane and, therefore, follow the membrane potential are called distributional probes. Depending on the charge of the probe (anionic or cationic) they accumulate in polarized or depolarized cells (Fig. 10a,b). Because of the diffusion times involved, these probes have slow (above seconds) response times. They can also be subject to active transport processes (pumps). Their biggest advantage is their high signal intensity change per millivolt membrane potential. With most probes capable of binding inside the cell, the cytoplasm reacts as a sink for the fluorochrome that has entered the cell, thereby removing it from the bulk of free diffusable dye molecules. This, in turn, allows even more molecules to accumulate inside the cell. If the extracellular dye concentrations are too high, all intracellular binding sites become saturated and the staining loses its dependency on membrane potential (see Fig. 10b,c) [20,34].

An excellent summary of membrane potential measurements and the function of the probes is given in Howard Shapiro's book *Practical Flow Cytometry* [10, pp.78–198]. He also describes the alternatives to distributional probes, that depend on the potential-driven electron distribution changes or orientation changes of dye molecules that intercalate into a membrane. The fast response of such probes to potential changes is offset by low signal intensity and signal change. Because of the signal variations between cells, these changes cannot be detected by flow cytometry, and the probes are suitable for only bulk measurements in cuvettes [20,34].

With the signal of distributional probes depending on a concentration equilibrium of free diffusable dye, signal intensity depends on membrane potential, but also on cell volume and the number of intracellular-binding sites available. As the latter two can change with growth conditions over time, slow changes of signal (>30 min) have to be treated with caution.

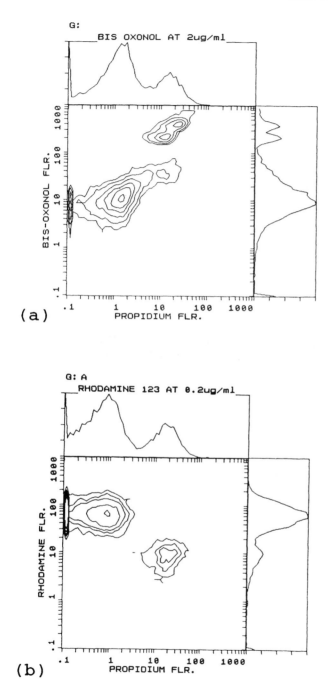

(a)

(b)

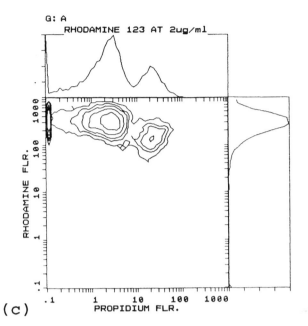

FIGURE 10 The effect of charge and concentration of membrane potential probes: To generate a truly depolarized population of cells, *Micrococcus luteus* that had been grown from lyophylized cells was mixed with heat-fixed cells from the same culture. Permeabilized cells are detected by propidium iodide fluorescence (*x*-axis) counterstained with (a) bis-oxonol and (b) rhodamine 123. The anionic bis-oxonol results in a very intensive fluorescence for the heat-fixed (depolarized) cells, making them appear red and green. The tail of propidium iodide-stained cells emerging from the unstained population were dead in the original culture. The cationic probe rhodamine 123 in accumulates in only polarized cells that are not permeabilized (propidium iodide-negative). (c) Staining with high concentrations of rhodamine 123 causes saturation of binding sites in intact and permeabilized cells, disrupting the detection of polarized cells.

Buffer systems used for staining have a profound influence on membrane potential itself and on the dye distribution obtained. This is demonstrated when looking at the Rhodamine 123 uptake of *Listeria innocua* (Table 3). Whereas starved cells show staining in all buffers, exponentially growing cells stained only in the salt-free buffer systems used. Apart from extracellular salt concentration, the energy provided to the cells will influence concentration equilibrium and active transport systems for the charged molecules including the dyes. Therefore, membrane potential estimations should be performed under the exact

TABLE 3 Comparison of Rhodamine 123 Uptake by Starved and Exponentially Growing *Listeria innocua* Cultures

Staining solution	Exponential cells	Starved cells
Tris buffer	+	+
Dist. water	+	+
0.2% glucose	+	+
TPB medium	–	+
PBS buffer	–	+
0.85% NaCl	–	+
0.85% KCl	–	+

conditions of interest. Detergents also have effects on the dye distribution by decreasing or increasing dye concentration in the medium, or by altering the membrane permeability of the dye.

Problems can occur when active transport systems remove the stain, as interference with the pump systems will also distort the membrane potential. Unless the problem can be solved by measuring depolarization, as with anionic probes, membrane potential measurements cannot be achieved.

Measurements of specific ion concentrations, such as calcium, have not yet been reported in bacteria, although they are now frequently performed or attempted in eukaryotic cells. Measurements with indo-1 or fura-2 might be possible, as they are ratiometric probes and, therefore, less sensitive to variability of intracellular dye concentrations (see next paragraph) *if* the straining is bright enough to generate a signal.

B. Intracellular pH

Maintenance and regulation of intracllular pH is no less important for bacteria than for eukaryotic cells. Consequently, its measurement is a valuable route to gaining insight into bacterial physiology.

In mammalian cells, direct pH measurement with micro-electrodes can be attempted for a single cell. It is certainly not an option for the much smaller bacterial cells. When looking beyond single-cell measurements to bulk cell studies, for any cell type, the method of choice is to use pH-sensitive fluorochromes. Similar to colorimetric indicators, their absorption and, in addition, their fluorescence spectra are sensitive to ionization. Fluorescein isothiocyanate, for example, when excited at 488 nm, loses nearly all its fluorescence when the pH drops from 8 to 6. The first description of intracellular pH measurement by flow cytometry [35] was based on an elegant trick, still favored today, for getting stain inside the cells. Retention of dye is favored by charge, but passage across cell membranes is favored by hydrophobicity and lack of charge. Thus, use of fluorescein diacetate allowed passage across the membrane, and then, intracellular esterases were used to cleave the charged fluorescein for retention. It was observed that the change in fluorescence intensity with pH was reversed when the fluorescein was excited above or below 465 nm. At this wavelength, the fluorescence intensity was pH-independent (isosbestic point). Thus, by measuring the ratio of fluorescence intensities taken at two wavelengths, one away from the isosbestic point and one at the isosbestic point, it is possible to have a parameter that is independent of the actual number of dye molecules present, either in the whole sample, or for flow cytometry, in each cell.

Either excitation or emission changes can be used for ratiometric measurements, depending on the fluorochrome used. Most fluorescein derivatives provide excitation ratios; thus, that usually means using two light sources. To a limited extent one of these (commonly abbreviated BCECF) can also be used for emission ratio measurements at 570 nm/520 nm [36]. A better probe for emission ratio measurements is the UV-excited 2,3-dicyanohydroquinone [37]. Companies, such as Molecular Probes, have derivatized numerous fluorescein compounds, such as those designated SNARF and SNAFL, to produce dyes with a variety of properties such as different excitation and emission wavelengths and emission and excitation ratios [33, 38]. The pH indicator with the lowest pK_a (4.2) is dicarboxydichlorofluorescein which allows measurements down to pH 4 [39].

Molenaar et al. [40] have shown, in *Lactococcus lactis*, some of the limitations of the pH probe BCECF caused by a lack of dye retention and a loss of the isosbestic (pH-independent) point, in their cuvette-based measurements. After loading the dye by acid shock, they showed good dye retention over 2.5 h only on ice. Moderate retention was observed at 30°C. Significant loss was observed when lactose, an energy source, was added to the cells. The addition of nigericin and valinomycin (which abolished ion gradients to allow calibration of the staining) caused no enhancement of loss, as determined by the appearance of the dye in the supernatant.

Despite such difficulties, there are still good chances for success. Dye loading at physiological pH using the diacetate or the acetoxymethylester of BCECF might cause less harm to bacteria, thereby enhancing retention. Loss of dye, whether active or passive, might also be minimized by the use of succinimidyl ester derivatives of the pH probes, as described under cell tracking (see Sec. VIII). This allows covalent coupling of dyes to the intracellular compartment, which minimizes dye losses. The loss of ratiometric behavior of the dye could be due to changes in autofluorescence with pH and might not be an issue for other cells or other probes, depending on their spectra.

C. Enzyme Activities

Enzyme activity has long been used as a measure for metabolic activity. Changes in enzyme systems are relatively slow compared with membrane potential changes and, thus, less prone to measurement artifacts. Therefore, stains can be loaded in buffer systems that facilitate dye loading and dye retention.

Esterase substrates are the most frequently used dyes to detect metabolic activity (see Chap. 15). They are presented as nonfluorescent acetoxymethyl ester (AM) or diacetate (DA) derivatives of fluorescent compounds, mostly fluorescein. After cleavage, they become fluorescent, polar, and charged; therefore, they are retained inside the cell, as discussed for pH probing.

There are several reasons why living cells may not stain with esterase substrates. First, the cells may not possess esterases, or the enzyme activity is extremely low owing to prestaining treatment or culture conditions. If that can be excluded the following three points have to be looked at:

1. Dye Loading

It is possible that the dye cannot enter the cell. Fluorescein diacetate (FDA) has, for example, been carboxylated to improve loading (CFDA). It is worthwhile trying different fluorochromes in their DA or AM form to test permeability. Another factor that can inhibit loading is extracellular cleavage because of enzymes or other active compounds present outside the cells, as well as charge owing to extracellular pH [10, p. 196; 33].

2. Dye Retention

As the dye is retained because of polarity and charge, the level of retention also depends on the membrane potential. Our investigations, in *L. innocua*, showed that incubation in Tris or distilled water improved the signal. This can be due to hyperpolarization or interference with pumps that might actively remove the dye. Very fast washing steps, such as washing on filter membranes, or sampling directly from the staining solution using high-dilution ratios, have to be used to detect fast-destaining cells.

3. Dye Quenching

If the fluorescence is quenched owing to intracellular pH, chlorinated fluorescein derivatives or calcein-AM should be used for esterase activity measurements. If cytoplasmic constituents interfere with absorption or emission it may be necessary to switch to a dye with different spectral properties.

Other fluorescent substrates, especially for use in molecular biology, have become available in the last few years; for example, to check for the functionality of genetic inserts using indicator enzymes, such as β-galactosidase [33, 41].

The tetrazolium salt CTC recently released by Polysciences looks to have significant potential for measuring dehydrogenase activity [14, 42]. It is a strong electron acceptor and precipitates as a red fluorescent crystal inside the cell. Unfortunately, the stain is very "temperamental," as the substrate can react with a variety extracellular components, and the staining requires specific conditions to work [43]. The ability to reduce tetrazolium depends on the oxidation–reduction potential of a bacterium. This variation in redox potential can turn tetrazolium salts into differential stains [44].

VI. SURFACE BINDING SITES AND GENERAL STAINING

By analogy with microscopy, the use of colored or fluorescent stains, of both surface materials and internal components, is a standard method for the characterization of all cells, including bacteria. The flow cytometer can use such stains, especially the fluorescent ones, for the same purpose.

Antigen–antibody recognition is the major application of flow cytometry in clinical immunology. There, white blood cells are differentiated, or functional epitopes or receptors are identified, by fluorescent-labeled antibodies. Thus, we might expect a similar situation for bacteria. Perhaps, owing to the much smaller size of most bacteria, compared with blood cells, and the consequently much higher demands made on instrumental sensitivity, antibody staining of bacteria is still, relatively, in its infancy. Now, however, the number of antibody-based bacterial identification assays is growing (e.g., Oxoid's latex agglutination tests for various bacteria and recent strip assays for *Listeria*), and flow cytometric-based studies are appearing more frequently to provide information about bacterial detection and differentiation [13,16,21,27,45–53]. Despite the smaller surface area of bacteria, compared with white blood cells, good separation between stained and unstained bacteria can sometimes be achieved (Fig. 11) [21]. Further problems arise because some antigens may be hidden from the antibody by other cell wall components [54], or capsular antigens may vary even within a strain [55]. Although, on one hand, variations in antigenicity by mutation, cell cycle, or growth may pose problems, they can also be used to identify changes in the bacterial culture. In addition, flow cytometry allows comparison of antigen density across a population of organisms. For use as a species or group detection agent, an antibody is of maximum use if it is known to bind to all subpopulations present (e.g., different stages of the growth cycle). Thus, the flow cytometer has been a powerful tool in the selection of new monoclonal antibodies for a variety of bacterial studies [G. Nebe-von Caron and R.A. Badley, unpublished studies, 1994].

If an antigenic determinant is the result of expression of a genetic construct, flow cytometry can be used to screen successfully, or even to sort and

G: B

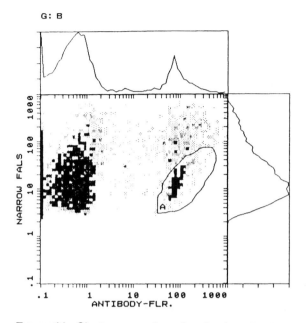

FIGURE 11 Cluster separation of antibody-stained bacteria from unstained cells in natural samples: Oral plaque was sonicated and stained with an IgM monoclonal antibody against a carbohydrate antigen of bacterial isolate *Streptococcus* 179. The antibody was indirectly labeled with monoclonal rat antimouse-kappa conjugated to phycoerythrin. Region A marks the single antibody-stained cells, but coaggregation with others cells is also present.

collect, on the basis of that expression. A recent example showed the successful enrichment of *Escherischia coli* expressing a single-chain Fv antibody fragment on its surface [56].

Analogously to antibodies, lectins can be used to gain information about the presence or absence of particular sugar residues on cell surfaces [57, 58]. One of the reasons for their relatively low popularity probably lies in the rather poorer affinity of a lectin for a target compared with typical antibody-antigen reactions. In addition, there are often difficulties with specificity and non-specific-binding reactions. However lectins may be worth considering in specific circumstances, since characterization of surface carbohydrates is difficult to obtain by other means.

Surface receptors can, in principle, be quantified by providing appropriate binding proteins, although there is still a dearth of examples with bacteria. Again, the small surface area and consequent low signals may be partly responsible.

Less specific than most antibodies, Gram staining of bacteria has been the most widely used primary characterization of all. It is included here because the primary difference between gram-positive and gram-negative bacteria lies in their cell wall compositions, even though some of the effects employed analytically are manifested elsewhere in the cell. Modern versions of this differential behavior are now available, including fluorescent versions. Molecular Probes has released a fluorescent Gram stain based on membrane-permeant nucleic acid stains. Staining with lipophilic dyes, such as carboxycyanines, in the absence and presence of EDTA, also allows determination of the percentage of gram-positive and grams-negative cells in a sample [20].

VII. NUCLEIC ACID, DNA PROBES, AND REPORTER GENES

Bacteria can be characterized by their nucleic acid. The first approach in this direction was taken by Van Dilla [59], who used differences in the A-T/G-C ratio of *Staphylococcus aureus*, *E. coli*, and *Pseudomonas aeruginosa*, stained with two DNA fluorochromes, with preferred binding to either of the base pairs. DNA measurements, combined with light scatter, allow separation of different-sized microorganisms in some cases [28]. Nucleic acid probes can target sequences of rRNA, and fluorescent labeling of the probe has allowed species identification in selected examples [48].

VIII. CELL CYCLE-BASED CHARACTERIZATION

The simplest way to show cell proliferation in normal batch cultures, one major consequence of cell cycle activity, is to count the number of bacteria. The statistical counting error can be kept very low by counting enough cells (see Sec. II) to allow the observation of the outgrowth of small subpopulations. Functional or differential stains allow even further differentiation.

Another approach, which is more geared toward the demonstration of heterogeneity in a population, is the tagging of cells with a covalent marker, sometimes called cell tracking. Figure 12 shows an overlay of a dot plot of starved *L. innocua* stained with the succinimidyl ester of dichlorocarboxyfluorescein before and after their first cell division. It is apparent that, at the second time point, not all cells have yet divided. Some cells remain highly fluorescent, but have increased light-scatter signals, compared with the initial cluster.

Difficulties arise when long lag phases or very long generation times occur, or when the cells do not physically separate. The change in light scatter that goes along with growth can sometimes be used to indicate proliferation. This can be due to a change in hydration, as in sporulation (see Fig. 5), or an increase in biomass, as in *Listeria* (see Fig. 12).

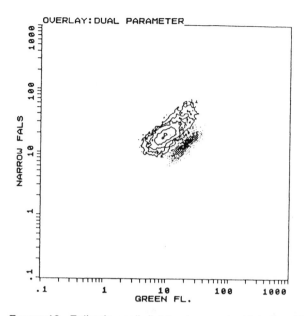

FIGURE 12 Following cell division by covalent labeling: Tracking of cell division of starved *Listeria innocua* by narrow forward-angle light scatter (*y*-axis) and covalent fluorescent labeling (*x*-axis) at 60 mins (dot plot) and 180 mins (contour plot). The dot plot represents a single cluster of undivided cells. The cells in the contour plot are well into cell division. They show one cluster with the same fluorescence as the dot plot, but a higher FALS signal and a second cluster of lower size and half the fluorescence intensity.

Quite a number of studies have investigated nucleic acid replication in cell growth [5,6,13,17,18,60–62]. The fundamental work has been done by Steen [7] who, using a DNA-specific staining mixture of mithramycin (plicamycin) and ethidium bromide, demonstrated that it is possible to show the patterns of DNA replication for bacteria by flow cytometry, in agreement with current growth models. He also showed these changes to be sensitive responses to antimicrobial reagents [7].

IX. VIABILITY

Probably the single most important characterization required by the practical flow cytometrist when working with "real"—nonlaboratory—samples is to be able to distinguish bacteria from the mass of particles usually present (e.g., in environmental water, sewage, and food samples; also see Chap. 16). This reduces down to whether a particle is, or has been, alive (viable). This important topic has been left until last because any or all of the bacterial characteristics already discussed can potentially be useful in defining the viability status of a particle seen in the flow cytometer. The complexity of the different meanings that "live" and "dead" can take on will be discussed, together with an indication of the types of flow cytometric measurements that can help in understanding and classifying the various states.

Life is characterized by

The presence of structure
Changeable genetic information allowing evolution
Metabolism or functional activity
An ability to reproduce or grow

Since the ability to reproduce has come to outweigh all the preceding definitions, it is the most stringent proof of life. The presence of *reproductive viable cells*, therefore, becomes the primary measure for viability. In bacteriology, the rapid cell division seen with most organisms can lead to massive amplification of cell numbers and the appearance of colonies of cells from a single starting cell.

Because of the long lag phases and cells' sensitivities to growth conditions, this definition presents a problem for investigation of stressed or injured cells. With single-cell investigation methods, such as flow cytometry, the discrepancy becomes apparent; there are more cells present than those that can grow and divide. Thus, the second definition, used for *vital cells*, is the presence of cellular metabolism (e.g., enzyme activities, oxidative reactions, and nucleic acid synthesis) [6, 14, 15, 25, 42, 63, 64].

In the absence of significant cell metabolism or cell division, single-cell measurements can still provide further methods of differentiation, by probing for the function of the most important barrier structure in the cell, the cell membrane, which has to protect the genetic information of the cell. *Intact cells* show selective permeability of the membrane. However, once the cell membrane integrity is lost and both metabolic activity and cell division are absent, detected events can be classified as *dead cells*.

Multicolor flow cytometry has provided a powerful method for detecting and analyzing these various states of viability of bacteria. The four states defined in the foregoing are summarized in Table 4.

A. Measurement of States of Viability

1. Proliferating or Reproductive Cells

The increase in cell numbers can be achieved under laboratory conditions and detected by classic methods. When bacteria thrive only in the presence of symbiotic partners, plate-counting methods and limiting dilution curves are bound to give false results. With the ability to discriminate bacteria from other particles

TABLE 4 The Four States of Viability

Proliferating or reproductive cells
Growth under laboratory conditions, using traditional plate culture methodology, can be detected.

Vital cells
Metabolic activity or cellular response to outside conditions can be demonstrated.

Intact cells
Selective permeability or cell wall integrity is detectable.

Dead cells
Particles with genetic information, but without an intact cell membrane, any signs of metabolic activity, or cellular response, or any indications of cell division or growth. As the cellular structures are freely exposed to the environment they inevitably decompose.

by nucleic acid stains, flow cytometry can be used to detect an increase in cell number simply by counting. In mixed cultures, differential staining might prove necessary for counting the different bacteria, but sometimes light scatter can be sufficient [65]. The halving of the amount of an added tracer within cells after division is another possible route to show cell division. Counting and tracking have already been discussed in their own right earlier (see Sec. VIII). When compared with all other methods, flow cytometry can provide proliferation information practically on-line with the first divisions.

2. Vital Cells

The demonstration of an energy-consuming or -releasing process is a growth-independent indication of cellular vitality. The major parameters that can be assessed by flow cytometry and that all have been discussed in the foregoing are

Enzyme functions
Membrane potential
pH regulation
Nucleic acid synthesis

The concept of cells being viable, but nonculturable, has become an important one, especially in relation to bacteria taken from nonlaboratory environments (see Chap. 16), but it is also important in laboratory or fermentation culturing. For example, the detection of a food pathogen that may have deposited significant amounts of a toxin before losing reproductive capability is likely to be a useful piece of information for the food manufacturer. The problem has also been identified for *Vibrio cholerae* in the environment [66]. The whole issue of dormancy has been reviewed by Kaprelyants et al. [67].

3. Intact Cells

Detection of intact cells is based on the demonstration of selective permeability or cell wall integrity, usually based on dye-retention or dye-exclusion measurements. The correlation of such cells with growth is difficult, and sometimes impossible, to achieve. The important feature of these cells is that they are still a separate entity from the environment and, therefore, have the potential to give rise to metabolism or proliferation. This is an important argument for risk assessment and more similar to the viability definition of viruses that also do not show direct proliferation or metabolism.

4. Dead Cells

A cell without an intact membrane cannot maintain any of the electrochemical gradients necessary to remain functional. Unless temporary permeability is deliberately caused (e.g., by electro- or chemoporation for inserting genetic constructs), such cells can be classified as dead. As all their structures are freely exposed to the environment, they will eventually decompose. This can be demonstrated by treatment with DNase and subsequent DNA staining. Membrane integrity can be measured by dye exclusion, using combinations of supravital and membrane-impermeant dyes, such as ethidium bromide and propidium iodide, or dye retention of fluorescent enzyme substrate stains (see Sec. IV).

5. Cell Damage

Viability measurements are central to all questions related to cell stress and injury. Membrane potential has been used as a measure of vitality by various authors and has been correlated with metabolic activity and antibiotic sensitivity. Loss of membrane potential indicates damage and can be achieved by stress or starvation or as a response to antibiotic challenge [63,64,68]. The fact that starved cells do show membrane potential staining after resuscitation [15] indicates its reversible character. Enzyme activity can also fall below detection level, thereby limiting the applicability of enzyme substrates. Lethal damage is difficult to show unless loss of cell wall integrity occurs. Unfortunately, this membrane breakdown does not necessarily occur immediately. Sometimes damage can, or has to be, demonstrated by increased sensitivity to toxic reagents or detergents [69].

X. CONCLUSIONS

A major conclusion that can be drawn, with some justification, from the material reviewed herein is that bacterial characterization by flow cytometric methods is now eminently possible, especially with the advent of newer commercial cytometers. As suggested in the introduction, two themes have emerged; namely, the benefits of multiparameter measurements, and the effect on sensitivity of the small size of most bacteria. Successful application of flow cytometry to bacteria, therefore does require some precautions to enhance the chances of success. These have been highlighted in the text, but frequently derive from flow cytometry's need for single cells (i.e., a lack of aggregation). Thus, flow cytometry creates a problem with this requirement, but also provides the solution in the form of multiparameter measurements and, indeed, once the problem has been overcome, additional information is available.

The application of flow cytometry to bacterial counting drew attention to the frequently observed discrepancy between counts obtained by classic culture techniques and those from flow cytometry. These differences can be a rich source of information about the viability patterns of cell cultures. In viability measurements, membrane integrity has been one of the safest ways to estimate cell life and death. Two techniques are really in their infancy for application to bacteria. Functional tests, for example, those metabolically based, are growing rapidly in scope and, by analogy with such tests in eukaryotic cells, will perhaps ultimately become some of the most useful. The application of antibodies has suffered from the lack of availability of numerous high-quality reagents. This situation is now changing, and flow cytometry itself can help in the selection of better antibody clones.

Finally, it is increasingly clear that flow cytometry has helped open a whole new set of definitions of bacterial life and death. The distinctions are not mere academic niceties, but can result in profound and significant practical consequences for medicine, food manufacture, and environmental monitoring. Viewing bacteria by flow cytometry has certainly provided exciting new ways for their characterization.

REFERENCES

1. F. T. Gucker, C. T. O'Konsi, et al., A photoelectronic counter for colloidal particles, *J. Am. Chem. Soc. 69*:2422 (1947).
2. R. M. Ferry, R. M. Farr, et al., The preparation and measurement of the concentration of dilute bacterial aerosols, *Chem. Rev. 44*:389 (1949).
3. A. S. Paau, J. R. Cowles, et al., Flow-microfluorometric analysis of *Escherichia coli, Rhizobium meliloti,* and *Rhizobium japonicum* at different stages of the growth cycle, *Can. J. Microbiol. 23*:1165 (1977).
4. M. L. Slater, S. O. Sharrow, et al., Cell cycle of *Saccharomyces crevisiae* in populations growing at different stages, *Proc. Natl. Acad. Sci. USA 74*:3850 (1977).
5. J. E. Bailey, J. Fazel-Madjlessi, et al., Characterization of bacterial growth by means of flow microfluorometry, *Science 198*:1175 (1977).

6. K. J. Hutter and H. E. Eipel, Flow cytometric determination of cellular substances in algae, bacteria, moulds and yeasts, *Antonie Leeuwenhock Microbiol. 44*:269 (1978).

7. H. B. Steen, Flow cytometric studies of microorganisms, *Flow Cytometry and Sorting*, 2nd ed. (M. R. Melamed, T. Lindmo, and M. L. Mendelsohn, eds.), Wiley-Liss, New York, 1990.

8. G. Frelat, C. Laplace-Builhe, et al., Microbial analysis by flow cytometry: present and future, *Flow Cytometry Advanced Research and Clinical Applications*, Vol. 2 (A. Yen, ed.), CRC Press, Boca Raton, FL, 1989, p. 275.

9. E. Boye and A. Lobner-Olesen, Flow cytometry: illuminating Microbiology, *New Biol. 2*:119 (1990).

10. H. M. Shapiro, *Practical Flow Cytometry*, 2nd ed., Alan R. Liss, New York, 1988.

11. D. B. Kell, H. M. Ryder, et al., Quantifying heterogeneity: flow cytometry of bacterial cultures, *Antonie Leeuwenhoek Microbiol. 60*:145 (1991).

12. D. Lloyd, ed., *Flow Cytometry in Microbiology*, Springer-Verlag, London, 1992.

13. H. Christensen, L. R. Bakken, et al., Soil bacterial DNA and biovolume profiles measured by flow cytometry, *FEMS Microbiol. Ecol. 102*:129 (1993).

14. A. S. Kaprelyants and D. B. Kell, The use of 5-cyano-2,3-ditolyltetrazolium chloride and flow cytometry for the visualisation of respiratory activity in individual cells of *Micrococcus luteus, J. Bicrobiol. Methods 17*:115 (1993).

15. A. S. Kaprelyants and D. B. Kell, Dormancy in stationary phase cultures of *Microcuccus luteus*: flow cytometric analysis of starvation and resuscitation, *Appl. Environ. Microbiol. 59*:3187 (1993).

16. J. Porter, C. Edwards, et al., Rapid, automated separation of specific bacteria from lakewater and sewage by flow cytometry, *Appl. Environ. Microbiol. 59*:3327 (1993).

17. B. R. Robertson and D. K. Button, Characterizing aquatic bacteria according to population, cell size and apparent DNA content by flow cytometry, *Cytometry 10*:70 (1989).

18. E. Boye and A. Lobner-Olesen, Bacterial growth control studied by flow cytometry, *Inst. Pasteur Elsevier 142*: 131 (1991).

19. B. Cantinieaux, P. Courtoy, et al., Accurate flow cytometric measurements of bacteria concentrations, *Pathobiology 61*:95 (1993).

20. H. M. Shapiro, *Flow Cytometric Approaches to Clinical Microbiology*, 2nd ed. Labmedica, 1990 p. 21.

21. A. Völsch, W. F. Nader, et al., Detection and analysis of two serotypes of ammonia-oxidizing bacteria in sewage plants by flow cytometry, *Appl. Environ. Microbiol. 56*:2430 (1990).

22. P. J. Wyatt, Differential light scattering: a physical method for identifying living bacterial cells, *Appl. Optics 7*:1879 (1968).

23. B. V. Bronk and W. P. Van De Merwe, In vivo measure of average bacterial size from a polarised light scattering function, *Cytometry 13*:155 (1992).

24. G. C. Salzman, S. B. Singham, et al., Light scattering and cytometry, *Flow Cytometry and Sorting*, 2nd ed. (M. R. Melamed, T. Lindmo, and M. L. Medelsohn, eds.), Wiley-Liss, New York, 1990, p. 81.

25. H. B. Steen and E. Boye, Applications of flow cytometry on bacteria: cell cycle kinetics, drug effects and quantitation of antibody binding, *Cytometry 2*:249 (1982).

26. R. Allmann, A. C. Hann, et al., Growth of *Azotobacter vinelandii* with correlation of Coulter cell size, flow cytometric parameters and ultrastructure, *Cytometry 11*:822 (1990).

27. L. A. van der Waaij and G. Mesander, Direct flow cytometry of anaerobic bacteria in human feces, *Cytometry 16*:270 (1994).

28. R. Allmann, A. C. Hann, et al., Characterisation of bacteria by multiparameter flow cytometry, *J. Appl. Microbiol. 73*:438 (1992).

29. G. J. B. Dubelaar, J. W. M. Visser, and M. Donze, Anomolous behaviour of forward and perpendicular light scattering of *Cyanobacterium* owing to intracellular gas vacuoles, *Cytometry 8*:405 (1987).

30. F. Scrienc, B. Arnold, and J. E. Bailey, Characterisation of intracellular accumulation of poly-ß-hydroxybutyrate (PHB) in individual cells of *Alcaligenes eutrophus* H16 by flow cytometry, *Biotechnol. Bioeng. 26*:982 (1984).

31. K. D. Wittrup, M. B. Mann, et al., Single cell light scatter as a probe of refractile body formation in recombinant *Escherichia coli*, *Biotechnology. 6*:423 (1988).

32. B. Lambert and J. B. Le Pecq, Effect of mutation, electric membrane potential and metabolic inhibitors on the accessibility of nucleic acids to ethidium bromide in *Escherichia coli* cells, *Biochemistry 23*:166 (1984).

33. *Handbook of Fluorescent Probes and Research Chemicals,* 5th ed., Molecular Probes Inc., Eugene, OR, 1992.

34. H. M. Shapiro, Flow cytometric probes of early events in cell activation, *Cytometry 1*:301 (1980).

35. J. W. M. Visser, A. A. M. Jongeling, et al., Intracellular pH-determination by fluorescence measurements, *J. Histochem. Cytochem. 27*: 27 (1979).

36. B. J. Hernlem and F. C. Scrienc, Intracellular pH in single *Saccharomyces cerevisiae* cells, *Biotechnol. Tech. 3*:79 (1989).

37. G. Valet, A. Raffael, et al., Fast intracellular pH determination in single cells by flow cytometry, *Naturwissenschaften 68*:265 (1981).

38. J. E. Whitaker, R. P. Haugland, et al., Spectral and photophysical studies of benzo[c]xanthene dyes: dual emission pH sensors, *Anal. Biochem. 194*:330 (1991).

39. M. Nedergaard, S. Desai, et al., Dicarboxy-dichlorofluorescein: a new fluorescent probe for measuring acidic intracellular pH, *Anal. Biochem. 187*:109 (1990).

40. D. Molenaar, T. Abee, et al., Continuous measurement of cytoplasmic pH in *Lactococcus lactis* with a fluorescent pH indicator, *Biochem. Biophys. Acta 1115*:75 (1991).

41. M. Manafi, W. Kneifel, et al., Fluorogenic and chromogenic substrates used in bacterial diagnostics, *Microbiol. Rev. 55*:335 (1991).

42. G. G. Rodriguez, D. Phippe, et al., Use of a fluorescent redox probe for direct visualisation of actively respiring bacteria, *Appl. Environ. Microbiol. 58*:1801 (1992).

43. S. M. Thom, R. W. Horobin, et al., Factors affecting the selection and use of tetrazolium salts as cytochemical indicators of microbial viability and activity, *J. Appl. Bacteriol. 74*:433 (1993).

44. E. M. Barnes., Tetrazolium reduction as a means of differentiating *Streptococcus faecalis* from *Streptococcus faecium*, *J. Gen. Microbiol. 14*:57 (1956).

45. C. W. Donnelly and G. J. Baigent, Method for flow cytometric detection of *Listeria monocytogenes* in milk, *Appl. Environ. Microbiol. 52*:689 (1986).

46. M. S. Obernesser, S. S. Socranske, et al., Limit of resolution of flow cytometry for the detection of selected bacterial species, *J. Dent. Res. 69*:1592 (1990).

47. B. K. Thorsen, O. Enger, et al., Long term starvation survival of *Yersinia ruckeri* at different salinities studied by microscopical and flow cytometric methods, *Appl. Environ. Microbiol. 58*:1624 (1992).

48. R. I. Amann, B. J. Binder, et al., Combination of 16s rRNA-targeted oligonucleotide probes with flow cytometry for analysing mixed microbial populations, *Appl. Environ. Microbiol. 56*:1919 (1990).

49. M. Van Hoegaerden, S. Levasseur, et al., Anti-*Enterobacteriaceae* common antigen (Anti-EGA) antibodies and their applications in specific detection and for the count of whole *Enterobacteriaceae* using an immunochemical method, *PCT Int. Appl.* (1992).

50. G. Vesey, G. Slade, et al., Taking the eye strain out of environmental *Cryptosporidium* analysis. *Lett. Appl. Microbiol. 13*:62 (1991).

51. K. J. Hutter, Simultane mehrparametrige durchflußzytometrische analyse verschiedener microorganismenspezies, *Monatsschr. Brauereiwissen. 9*:281 (1992).

52. J. E. Reseland, K. S. Cudjoe, et al., Use of flow cytometry for detection and quantification of *Clostridium perfringens* type A enterotoxin positive spores, *Food Safety Quality Assurance-Proceedings 1st Applied Immunoassay Systems*, 1992, p. 315.

53. J. M. Barnett, M. A. Cuchens, et al., Automated immunofluorescencent speciation of oral bacteria using flow cytometry, *J. Dent. Res. 63*:1040 (1984).

54. A. T. Bentley and P. E. Klebba, Effect of lipopolysaccharide structure on reactivity of anti-porin monoclonal antibodies with the bacterial surface, *J. Bacteriol. 170*:1063 (1988).

55. D. A. Lutton, S. Patrick, et al., Flow cytometric analysis of within strain variation in polysaccharide expression in *Bacteroides fragilis* by use of murine monoclonal antibodies, *J. Med. Microbiol. 35*:229 (1991).

56. J. A. Francisco, R. Campbell, et al., Production and fluorescence-activated cell sorting of *Escherichia coli* expressing a functional antibody fragment on the external surface, *Proc. Natl. Acad. Sci. USA 90*:1044 (1993).

57. J. A. De Stephano, L. S. Trickle, et al, Flow cytometric analysis of lectins binding to *Pneumocystis carinii* surface carbohydrates, *J. Parasitol. 78*:271 (1992).

58. R. A. Bloodgood, N. L., Salomonsky, et al., Use of carbohydrate probes in conjunction with fluorescence activated cell sorting to select mutant cell lines on *Chlamydomonas* with defects in cell surface glycoproteins, *Exp. Cell Res. 173*:572 (1987).

59. M. A. Van Dilla, R. G. Langlois, et al., Bacterial characterization by flow cytometry, *Science 220*:620 (1983).

60. J. Fazel-Madjlessi, J. E. Bailey, et al., Flow microfluorometry measurements of multicomponent cell composition during batch bacterial growth, *Biotechnol Bioeng. 22*:457 (1980).

61. K. J. Hutter and H. Oldiges, Alterations of proliferating microorganisms by flow cytometric measurements after heavy metal intoxication, *Ecotoxicol. Environ. Safety 4*:57 (1980).

62. U. von Freiersleben and K. V. Rasmussen, DNA replication in *Escherichia coli* gyrB(Ts) mutants analysed by flow cytometry, *Res. Microbiol. 142*:223 (1991).

63. J. P. Diaper, K. Tither, et al., Rapid assesment of bacterial viability by flow cytometry, *Appl. Microbiol. Biotechnol. 38*:268 (1992).

64. A. S. Kaprelyants and D. B. Kell, Rapid assessment of bacterial viability and vitality by rhodamine 123 and flow cytometry, *J. Appl. Bacteriol. 72*:410 (1992).

65. Y. Hechard, C. Jayat, et al., On-line visualisation of the competitive behaviour of antagonistic bacteria, *Appl. Environ. Microbiol. 58*:3784 (1992).

66. R. R. Colwell, B. R. Brayton, et al., Viable but non-culturable *Vibrio cholerae* and related pathogens in the environment: implications for the release of genetically engineered microorganisms, *Biotechnology 3*:817 (1985).

67. A. S. Kaprelyants, J. C. Gottschal, et al., Dormancy in non-sporulating bacteria, *FEMS Microbiol. Rev. 104*:271 (1993).

68. H. Bercovier, M. Resnick, et al., Rapid method for testing drug susceptibility of mycobacteria spp and gram-positive bacteria using rhodamine 123 and fluorescein diacetate, *J. Microbiol. Methods 7*:139 (1987).

69. B. Ray, Methods to detect stressed microorganisms, *J. Food Protect. 42*:346 (1979).

16

Assessment of Viability of Bacteria by Flow Cytometry

CLIVE EDWARDS
University of Liverpool, Liverpool, England

I. INTRODUCTION

Microbiology has long relied on traditional methods for measuring the numbers of bacteria present in any given sample. Preeminent in the development of the subject has been the plate count, whereby it is assumed that a viable cell exposed to the appropriate solid agar medium will, by rounds of growth and division, give rise to a single colony. Each colony is presumed to have arisen from a single cell; therefore, enumeration of the numbers of colonies that grow from suitably diluted samples gives the total numbers of viable cells in the original material. However, this principle, which is handed down to succeeding generations of microbiologists, is probably a gross oversimplification that has arisen largely as a result of the dominance of laboratory-based studies that employ rich growth media and ideal culture conditions. Such cultures yield cells that may be grossly different, both morphologically and biochemically, from those found in their natural environments. Several workers have remarked on the freakish nature of laboratory-grown cells that results in the growth of what can be viewed as giant cells, compared with those found in their natural habitats [35,47].

Natural environments are invariably severely nutrient-limited [24,25], and bacterial species exhibit an array of responses and properties that can often result in cells that are not detectable by traditional methods. To all intents and purposes, the traditional methods relied on by microbiologists fail, and many bacterial species in natural environments are not detected. It is from recent work in microbial ecology that our traditional definition of a viable cell has been challenged. As we shall see, any situation that leads to nutrient limitation grossly affects the properties of bacteria. This chapter will describe why rapid methods that can unequivocably identify and enumerate viable cells in the absence of culture methods are required; the responses that can be elicited by bacteria that make them necessary, and how flow cytometric methods (FCM) are being developed that herald rapid, and automated analytical methods. Other applications of flow cytometry in microbiology have also been recently reviewed [see 20, 22, 32].

II. THE NEED FOR METHODS THAT ASSESS BACTERIAL VIABILITY

Table 1 summarizes some of the areas for which quantitative enumeration of viable cells is desirable. Much of the impetus comes from studies of bacteria in natural environments, particularly marine ecosystems, which are almost always nutrient-limited [24,25,33]. A major feature of natural environments is that only a small proportion of the bacteria present can be cultured by traditional methods (57). This can range from 0.1 to 10%, which means that the true bacterial diversity of many habitats is unrealized. The reasons for this are complex. Part of the problem is our ignorance of the nutritional requirements of many species. Another major difficulty centers on what proportion of the uncultured fraction is alive and what is the true magnitude of the dead population. Recent work has shed some light on these problems by revealing that bacteria can exist in a non growing, but viable, state that has profound effects on their properties.

Several terms have arisen in the literature to describe bacteria in this state and these include ultramicrobacteria or dwarf cells (*Vibrio*) [see 25,31,47]; viable but noncultivable (VBNC) cells [12,41,55,60]; starvation-survival states [see 47]; dormant cells (e.g., *Micrococcus*), [see 30]. These terms represent

TABLE 1 Areas That Require Methods for Assessing the Viability of Bacteria

Natural environments	Water treatment
Detection of pathogens	Quality control
Release of GMMOs	Testing of antimicrobials
Commercial inocula	Commercial-scale
Bioremediation	fermentations

attempts to classify experimental observations. However, as more information accrues, it is probable that, in fact, these terms describe different facets of the same stress-induced process or response. The complex reactions that may occur when nondifferentiating bacteria are subjected to nutrient limitation, or to the severe starvation conditions that are typical of nearly all natural environments, result in sequential or overlapping states undergone by a bacterial cell during transit from a living vital form to death. It is important to emphasize that not all species need display exactly the same responses or have identical rates of progress through the transition from life to death. Some bacterial species are more long-lived under conditions of nutrient limitation than are others. For example, flow cytometric analysis of the viability of *Klebsiella pneumoniae* in lake water microcosms revealed that survival was much more prolonged than that observed for *Staphylococcus aureus* studied under similar conditions [14,15]. This illustrates why, in the literature so many different terminologies have originated that describe the physiological status of bacteria in natural environments or those exposed to nutrient stress. Some species will remain resuscitable using isolation on general growth media for most of the time, whereas others may display a transient or long-lived VBNC state. Marine species are probably the most adapted species to long-term survival in nutrient-limited conditions, and their responses have been extensively studied [33,34,54,55]. They represent those species that respond to carbon starvation by the formation of a small, metabolically inactive cell and exhibit a degree of differentiation, in that new structures and properties become apparent [51]. For other species, responses to starvation consist of a transient VBNC state before complete loss of viability, whereas other species merely display a gradual dying-off of cells with time (e.g., *Staphylococcus aureus* [14]).

Much of our understanding of starvation responses comes from laboratory-based studies that employ microcosms representative, to a greater or lesser degree, of a natural habitat, and it is important to issue a note of caution concerning their use. They often mimic natural environments poorly because they are closed systems—particularly aquatic microcosms—and as such allow the possibility of recycling processes whereby a low population of cells is permanently established that cycles through living and dead forms by cryptic growth. Such a situation need not arise in the more open, natural environments, where at localized sites such cycling through cryptic growth may not occur. In evolutionary terms the species we see now in soil, waters, and other natural habitats represent those species that have become highly evolved to adapt to long periods of starvation. Therefore, the ability to gauge the true incidence of viable bacterial numbers is important to many disciplines of microbiology.

Major areas of interest are listed in Table 1 and these include recreational and bathing waters, as well as any organization interested in investigating the true bacterial diversity of natural habitats (eg., screening programs of large pharmaceutical companies). Other examples are the need to rapidly identify viable pathogenic species, important to many diverse areas that include medicine, environmental health, and the food industry. Many environmental stimuli, including nutrient stress, have been recognized as regulators or effectors of pathogenicity in bacteria [45], and the tendency of many pathogenic Gram-negative bacterial species to become dormant and noncultivable has been high-lighted [41]. The development of rapid methods, such as FCM, for unequivoca-ble identification and enumeration of viable cells is, therefore, of great impor-tance, because the inability to culture a pathogen from a natural environment cannot now be taken as evidence for its absence. This is illustrated by work on pathogenic vibrios. *Vibrio cholerae* can become undetectable in aquatic envi-ronments, as assessed by traditional culture methods. However, the same water samples are known to be able to cause cholera in experimental animal models, implying that a VBNC state may be normal for this bacterium [12,60]. Similar observations have been made for the human pathogen *Vibrio vulnificus* [6,54], although more recent work has questioned the existence of a VBNC state in this bacterium [23,67]. The potential of pathogenic bacteria to exist in dormant states is of some significance in understanding the epidemiology of bacterially acquired diseases.

Recently, there has been a great deal of interest in the possible accidental or deliberate release of genetically modified microorganisms (GMMOs). Assessing how long such organisms will remain viable and active in open envi-ronments is of prime concern [see Ref. 19 for review]. Examples of such situa-tions are again given in Table 1. Commercial inocula that comprise a cocktail of genetically modified species have been advanced for a variety of purposes, particularly in the agricultural area; for example, those tailored specifically for improved silage production, better composting of plant matter, or stimulation of plant growth through improved nutrient supply (e.g., nitrogen provision in legu-minous and other plants by GMMOs for improved productivity and soil fertil-ity). Recently, bioremediation has been advanced as an area of intense interest by which bacteria, including genetically modified forms, may be released to mediate a particular function, such as cleanup of toxic spills, nuclear waste, and enhanced oil recovery from deep sea wells. Knowledge of the survival and viability of bacteria used in these ways is of prime importance. However, survival in natural environments is by no means predictable by traditional detec-tion methods. Other areas include measurements of viable numbers of organisms in drinking and other waters, as well as in food products. For these processes, it is desirable that methods are rapid and automated.

Another important area is that of testing antimicrobial compounds, for which a major difficulty is deciding whether an unknown compound is bacterio-static or bactericidal. Some applications of FCM in the field of drug testing have been described [18], but only relative to the effects of drugs on bacterial size and ploidy. Therefore, a rapid method of assessing viability in the absence of culturing methods, which may fail to detect damaged, but potentially viable, cells, would be highly desirable. Such a method would also be of benefit to commercial-scale fermentations, particularly those that result in very high product yield by using dense cultures. This problem is further complicated in processes in which products are synthesized as secondary metabolites. These are generally produced during the nongrowing stationary phase of growth at a point when cells are also beginning to die. Recently, the properties of bacteria in stationary-phase cultures has received a great deal of interest. For example, work on stationary-phase cultures of *E. coli* has demonstrated, in a proportion of the cells, prolonged survival that is accompanied by important biochemical and genetic events. These include the formation of mutant cells that display improved survival properties, as well as the capacity for higher competitiveness when introduced into nutrient-rich conditions. Their competitiveness is such that they have been reported to outgrow young early log-phase cultures [35,62,68]. This increased physiological fitness that arises because of starvation has important implications for all areas of microbiology.

In conclusion, the requirement for rapid, reproducible methods that clearly distinguish live cells, irrespective of their metabolic state or capacity for resuscitation or cultivation, is paramount and impinges on many important areas of microbiology

III. THE PROBLEMS OF ENUMERATING LIVE CELLS

For some of the examples in Table 1 bacteria may be in high-nutrient environments and, in this situation, determining colony-forming units (cfu) may be adequate, although this may be time-consuming and labor-intensive. Most of the areas listed in Table 1 have a strong possibility that the cells under investigation will be in environments in which nutrients are unavailable or severely limited. For nondifferentiating heterotrophs, starvation has many consequences, and the main one discussed in the foregoing is the failure of plating and other methods to detect all viable cells. Some of the reasons for this are given in Table 2.

The responses of bacteria to nutrient limitation have been known for a long time and stem from the elegant studies of Ole Maaloe and co-workers in the 1950s [see Ref. 61]. This work on nutritional shifts up and down as well as growth rate studies in nutrient-limited chemostats laid the foundations for understanding bacterial growth physiology and has often been overlooked by many microbial ecologists. Many of the responses to starvation listed in Table 2, such as formation of small cells that have reduced RNA and protein content, have been long known; however, environmental studies have enhanced our understanding of many of these changes. For example, the extent to which the size of

TABLE 2 Some of the Responses of Bacteria to Starvation

Response	Examples	Ref.
Reductive division: ultramicrobacteria	Soil bacteria Marine bacteria *Escherichia coli*	43 44 11,45
Protein turnover: synthesis of starvation- induced (sti) proteins	*Salmonella typhimurium* *E. coli* *Vibrio* S14	46 47 21,41
Reduction in total cellular RNA	Marine *Vibrio* spp. Numerous examples	48 49
Long-lived mRNA molecules	*Vibrio* S14	31
DNA levels remain constant or increase	Numerous examples	49
Reduced metabolic activity	*Vibrio* strain ANT 300	50,51
Morphological changes	*Vibrio* strain S14	21,32
Altered physiology: development of new resistance properties	*Vibrio* strain S14 *E. coli* Various examples	21,32 1 2
Changed antigenicity: new surface structures synthesized	*Vibrio* *Escherichia, Salmonella*	52 1
Cells may become viable but non- cultivable or dormant	Numerous examples, particularly gram- negative pathogens	11–13,15
Mutations in *rpoS*: cells more competitive for resuscitation and survival	*E. coli*	29

starving cells can be reduced, resulting in the formation of ultramicrobacteria, has been appreciated only through ecological studies. Starvation responses have been likened to a program of stress-induced differentiation, and this view stems largely from work with *Vibrio* spp. [2,34,51,52]. A major feature of these responses is the formation of a cell that is physiologically altered and frequently cannot be resuscitated by growth on media that normally support growth. Such cells have been termed viable, but noncultivable, but recently Kell and co-workers have adopted the term *dormant cells* [30,31], and this term will be used throughout this chapter. The central problem is to develop methods that accurately identify and allow enumeration of bacteria, irrespective of whether they can be cultured.

IV. FLUORESCENT DYES USED TO IDENTIFY VIABLE BACTERIA

Table 3 lists various fluorescent dyes that have been used in attempts to identify viable bacteria, and many of these have been used in conjunction with flow cytometry. Some have been especially useful for discriminating viable cells. Dyes are available that are actively accumulated into the cell in response to the transmembrane potential ($\Delta\varphi$)—a property of live cells only. One of these, the cyanine dye has very limited applications for bacteria, owing to a great deal of indiscriminate staining that occurs in dead as well as live cells. However rhodamine 123 (R123) has been highly effective for monitoring viability in a whole range of bacterial species, although it is by no means universally applicable. It has been used in conjunction with FCM to study survival and dormancy in *M. luteus* [28,30] and in survival studies in *Staphylococcus aureus* [14]. Its applicability for studying viability of bacteria has been discussed by Diaper and Edwards [16], and protocols for its use in the flow cytometric assessment of viability of both Gram-negative and Gram-positive species have been described [58]. For the latter, the dye is readily taken up by most species; however, Gram-negative bacteria need treatment with EDTA to allow the dye to pass through the outer membrane. Oxonol dyes are also membrane potential-dependent, but differ from R123 in that they are useful for staining dead cells. Normally the dye enters the cell and is immediately pumped out by a membrane potential-dependent mechanism; live cells, therefore, are not stained, but dead cells will retain the dye because they do not possess a $\Delta\varphi$. We have tested this dye in live–dead mixtures of a range of bacteria and found that FCM can be used to accurately enumerate the dead cells.

Colorless dye conjugates are cleaved enzymically inside a cell to release a fluorescent molecule that is retained and stains only live cells that have intact membranes, whereas dead cells, with damaged or leaky membranes, will extrude the dye. Several dye conjugates are listed in Table 3 some of which (FDA, CFDA, BCECF-AM, and calcein) are cleaved intracellularly by nonspecific esterases. FDA is probably of little use for the flow cytometric assessment of viability in bacteria. Our studies have shown that, although most bacteria possess the requisite enzymes to cleave FDA, the released fluorescein is pumped out of the cell at rates too rapid to allow staining of the cell [16,20]. Derivatives of FDA have become available, for which dye retention after enzymic cleavage is

TABLE 3 Some Fluorescent Dyes That Have Been Used to Assess Microbial Viability

Dye	Mode of action	Applications	Ref.
Dihexyloxacarbocyanine	Membrane potential	Flow cytometric (FCM) detection of bacteria	34,53
Rhodamine 123 (R123)	Membrane potential	Microscopic enumeration of viable bacteria	54,55
		FCM assessment of viability in	
		A range of gram-positive and	
		gram-negative bacteria	34
		Micrococcus luteus	33
		Staphylococcus aureus	17
Bis-(1,3-dibutylbarbituric acid) pentamethine oxonol (Oxonol)	Membrane potential	Enumeration of dead cells	Deere and Edwards, unpublished work
Fluorescein diacetate (FDA)	Intracellular esterase cleavage to release fluorescein, which is retained within cells possessing an intact membrane	Microscopic detection of	
		Viable gram-positive bacteria	56
		Viable mycobacteria	54,57
		Viable soil bacteria	58
		Viable aquatic bacteria	59
		FCM enumeration of	
		Metabolically active marine microalgae	60
		Viable *Bacillus subtilis*	18

Carboxyfluorescein diacetate (CFDA)	As for FDA	FCM enumeration and identification of viable compost bacteria	18
		FCM assessment of viability in *Saccharomyces cerevisiae*	39
2',7'-Bis-(2-carboxyethyl)5(6)-carboxyfluoresceinacetoxymethylester (BCECF-AM)	As for FDA	FCM assessment of viability of a range of bacterial species	37
Calcein acetoxymethylester	As for FDA	Microscopic detection of viable protozoa	61
		FCM assessment of viability of a range of bacterial species	37
Fluorescein di-β-D-galactopyranoside (FGP)	Intracellular enzymic cleavage, dye retention only by cells with intact membrane	FCM–activated cell sorting of viable yeasts and bacteria	62
Chemchrome Y	As for FGP	FCM detection of *Candida albicans*	63
Chemchrome B	As for FGP	FCM enumeration of viability of a range of bacterial species	17,37,64
		Analysis of viability of genetically modified *Bacillus subtilis* in compost	17
		Resuscitation of VBNC *Vibrio vulnificus*	24

TABLE 3 (continued)

Dye	Mode of action	Applications	Ref.
5-Cyano-2,3-ditolyl tetrazolium chloride (CTC)	Respiratory activity	Microscopic detection of active aquatic bacteria	65
		FCM detection of respiring *M. luteus*	38
		Analysis of dormancy in *M. luteus*	16
Plicamycin (mithramycin)	Staining elongated cells (viable) after prolonged incubation in the presence of nalidixic acid	FCM monitoring of viable *Yersinia ruckeri*	66
Propidium iodide	Dye exclusion by live cells	Microscopic detection of viable protozoa and yeast	67
4′,6-Diamidino-2-phenylindole (DAPI)	Dye exclusion	Microscopic detection of *Cryptosporidium parvum* oocysts	67

much improved compared with FDA. Of these CFDA and BCECF-AM are probably the best and are less susceptible to bleaching [48]. Other dyes have had more limited applications and include calcein-AM and FGP. Recently developed dyes for which the mode of action is unknown include chemchrome Y for assessing viability of yeasts and fungi and chemchrome B for bacteria. Chemchrome B has been especially useful for monitoring viability of a wide range of bacterial species [13] and has also been effective for FCM identification of live cells in a variety of microcosms representative of aquatic and particulate environments [15,23]. Recently, the use of CTC, which fluoresces in actively respiring cells has also been reported to be suited to the flow cytometric determination of viability in *M. luteus* [29].

V. APPLICATIONS OF VIABILITY DYES IN CONJUNCTION WITH FLOW CYTOMETRIC METHODS

For bacteria, membrane potential-dependent fluorescent dyes, notably rhodamine 123 and the dye conjugates, have been most widely studied for the assessment of viability in bacteria. Kaprelyants and Kell [28] examined the suitability of R123 for rapid assessment of the viability of *M. luteus* by flow cytometry. They confirmed that this dye entered the cell in response to the cell's membrane potential and showed active accumulation of the dye, such that at 0.87 μM of Rh123, $[R123]_{in}/[R123]_{out}$ was 3000:1. Furthermore dye uptake at these low concentrations could be completely abolished by the uncoupling agent carbonylcyanide-*m*-chlorophenylhydrazone (CCCP). This method was used to distinguish viable from dead cells using flow cytometry and to examine the viability profiles of a chemostat population of *M. luteus*.

Diaper and Edwards [16] extended these studies to examine the applicability of R123 as an indicator of bacterial viability. A range of species were examined with varying results. Viable cells of most bacterial species examined could be detected by FCM after staining with R123. Pseudomonads were particularly difficult to detect using rhodamine staining (Table 4). There is no doubt that this dye is highly suited to viability studies of pure cultures of those bacteria that can actively concentrate it inside the cell. Subsequent work has used this fluorochrome to examine the nature of dormancy in *M. luteus* and to monitor the resuscitation capacity of dormant cells [30,31]. However R123 is unsuited for studies of viability when bacteria are found in particulate habitats, such as those in natural environments or foods, for example. The colorless dye conjugates, which are cleaved inside the cell to release a fluorescent molecule retained by cells with intact membranes (therefore, viable), are far better suited to such a purpose. Some of these are listed in Table 3 and their success in identifying viable bacteria by FCM is summarized for a range of species in Table 4.

TABLE 4 Applicability of Various Dyes for Assessing the Viability of Selected Bacterial Species

Bacteria	Dyes				
	R123	CFDA	BCECF-AM	Calcein-AM	Chemchrome B
Staphylococcus aureus	+	+	+	+	+
Escherichia coli	+	+	+	−	+
Salmonella pullorum	+	−	−	−	+
Pseudomonas fluorescens	−	−	+/−a	−	+
Aeromonas hydrophila	+	−	−	−	+
Arthrobacter globiformis	+	−	+	−	+
Bacillus subtilis	+	+	+	+	+
Pseudomonas aeruginosa	−	−	−	−	−

aOnly partial staining of about 20% of total cells.

It is immediately apparent that most have very limited applications, especially calcein (see Table 4). However chemchrome B can be used to stain viable cells of most bacterial species that we have examined, apart from some notable exceptions preponderantly found in the genus *Pseudomonas*. Just why the pseudomonads appear to be recalcitrant to a variety of viability dyes remains unknown. Although the mode of action of chemchrome B has not been supplied by its manufacturers, it probably works in much the same way as the many FDA derivatives listed in Table 3 that are cleaved by nonspecific esterases once inside the cell. Chemchrome B may well have an alternative enzyme cleavage mechanism but, whatever this is, it is bound to involve enzymes that are common to most species. Of the dye conjugates listed in Table 4, the FDA derivatives are most successful with gram-positive species, and CFDA (as well as chemchrome B) has been used to monitor survival of a genetically manipulated *Bacillus subtilis* in compost, as well as to enumerate the total viable bacterial numbers present in what is a highly particulate environment [15]. More recently, CFDA has been shown to be a good indicator of viability in *Saccharomyces cerevisiae* [7]. Experimental methods for staining bacteria with dye conjugates have been described by us previously [58].

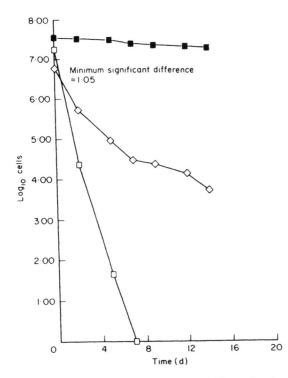

FIGURE 1 Stained cells of *Vibrio vulnificus* after incubation in sterile seawater at 4°C. Cell numbers of acredine orange-stained cells (■) remained constant for 16 days, but colony forming units (◇) tell rapidly. FCM enumeration of chemochrome B-stained cells (□) fell less rapidly, indicating that not all cells were dead.

The ability of FCM to enumerate and discriminate viable bacteria has been tested for *V. vulnificus* [23]. This bacterium is found in coastal waters around the world closely associated with oysters and other shellfish and has been identified as a human pathogen [54]. It is readily isolated from warm waters by plate counts on selective media, but this is not true in colder waters, and it has been suggested as due to a cold-mediated entry of cells into a VBNC state [53,56]. Figure 1 illustrates the problem whereby *V. vulnificus* was incubated in sterile seawater at 4°C. Total numbers estimated by microscopic enumeration of acridine orange-stained cells remained relatively constant for 16 days, but colony-forming units fell rapidly, such that no bacteria could be detected by this method after 8 days. Enumeration of chemchrome B-stained bacteria by FCM showed a decrease in numbers that was less rapid, indicating that not all cells were dead, even up to and including day 14. This experiment shows the diffi-

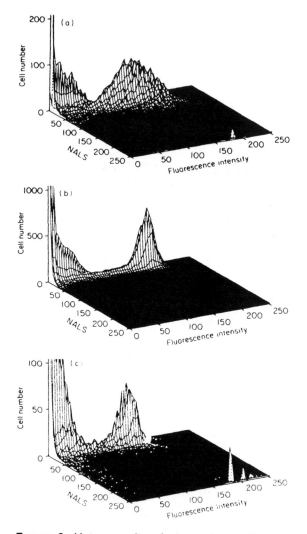

FIGURE 2 Heterogeneity of chemachrome B staining of unstarved *Klebsiella pneumonie* cells and its reduction in starved cells.

culty of interpretation of viable numbers enumerated by plate counts and the potential of FCM as a rapid diagnostic tool.

Chemchrome B has also been used to monitor the survival of *Klebsiella pneumoniae* in a lake water microcosm and, as with *V. vulnificus*, it was demonstrated that cell numbers obtained by FCM enumeration of stained cells were consistently higher than those from colony-forming units, suggesting that this organism can adopt a dormant state in which it is viable, but fails to grow on solid media normally used for its culture [13]. Another interesting observation made from this experiment was the heterogeneity of staining using chemchrome B that was observable in unstarved cells before inoculation into the microcosm and its subsequent reduction during survival in lake water. This is shown in Figure 2 in which unstarved cells show a wide variance in fluorescence intensities, such that cells possessing similar light scatter (therefore, are of similar size) display a range of fluorescence channel values. This heterogeneity is dramatically reduced in starved cells. Such an analysis is possible only through the use of FCM, which measures the fluorescent intensity of staining for each cell within a population. By reference to the light-scatter profile, it is possible to analyze the distribution of dye intensity in cells of different age (i.e., at different stages of the cell cycle). This is useful, because bacteria are generally morphologically uniform, with few cell cycle landmarks that can be visualized other than increase in size. This aspect has been discussed in two thoughtful articles that assess the applications of FCM for studying heterogeneity within bacterial populations [32,39].

VI. SUMMARY

In this chapter an attempt has been made to describe the difficulties that microbiologists now confront in describing whether a bacterium is viable or dead. The importance of this cannot be overstressed, particularly in the context of pathogenic species. There are now several methods, based on fluorescent molecules, to identify a viable bacterium, irrespective of whether it can be cultured. Many of these are also suited to analysis by FCM. There is no doubt that flow cytometric techniques have several advantages that make further studies in this area attractive. These include rapid enumeration; the lack of subjectivity that may be encountered in microscopic methods; the analysis of thousands of cells in seconds, so that the status of a whole population can be assessed; the methods can be automated; the data can be manipulated by computer software; a multiparameter approach is possible, whereby viability studies of heterogeneous populations can be linked to fluorescent probe analysis (e.g., antibody) for identification purposes; the FCM can be used to analyze heterogeneity within a population so that, for example, the spread of vitality within a bacterial population can be analyzed relative to the cell cycle. There are also some disadvantages, and for microbiologists, the major one would appear to be the problem of using FCM to analyze bacteria that inhabit particulate substrates for which some extraction process is necessary before passage of the cells through the cytometer. Such a step may itself affect viability of the sample under study. However, the techniques described in this chapter represent recent developments in the analysis of viability in bacteria, and there is no doubt that future progress will be made in this area which, as has been described in this chapter, will have wide-ranging applications for microbiology.

ACKNOWLEDGMENTS

This work was funded by grants from the NERC through the Tiger programme and through award GR3/7596.

REFERENCES

1. N. H. Albertson, G. W. Jones, and S. Kjelleberg, The detection of starvation-specific antigens in two marine bacteria, *J. Gen. Microbiol. 133*:2225–2231 (1987).
2. N. H. Albertson, T. Nystrom and S. Kjelleberg, Functional mRNA half lives in the marine *Vibrio* sp. S14 during starvation and recovery, *J. Gen. Microbiol. 136*:2195–2199 (1990).
3. L. Bakken and R. Olsen, The relationship between cell size and viability of soil bacteria, *Microb. Ecol. 13*:103–114 (1987).
4. H. Bercovier, M. Resnick, D. Kornitzer, and L. Levy, Rapid method for testing drug-susceptibility of *mycobacteria* spp. and gram-positive bacteria using rhodamine 123 and fluorescein diacetate, *J. Microbiol. Methods 7*:139–142 (1987).
5. M. Brailsford and S. Gatley, Rapid analysis of microorganisms using flow cytometry, *Flow Cytometry in Microbiology* (D. Lloyd ed.), Springer Verlag, London, 1993, pp. 171–180.
6. L. A. Brauns, M. C. Hudson, and J. D. Oliver, Use of the polymerase chain reaction in detection of culturable and non-culturable *Vibrio vulnificus* cells, *Appl. Environ. Microbiol. 57*: 2651–2655 (1991).
7. P. Breeuwer, J.-L. Drocourt, F. M. Rombouts and T. Abee, Energy-dependent, carrier-mediated extrusion of carboxyfluorescein from *Saccharomyces cerevisiae* allow rapid assessment of cell viability by flow cytometry, *Appl. Environ, Microbiol. 60*:1467–1472 (1994).
8. G. Brunius, Technical aspects of the use of 3'6'-diacetyl fluorescein for vital fluorescent staining of bacteria, *Curr. Microbiol. 4*:321–323 (1980).
9. A. T. Campbell, L. J. Robertson and H. V. Smith, Viability of *Cryptospiridium parvum* oocysts: correlation of *in vitro* excystation with inclusion or exclusion of fluorogenic vital dyes, *Appl. Envron. Microbiol. 58*:3488–3493 (1992).
10. E. A. Carter, R. E. Paul and P. A. Hunter, Cytometric evaluation of antifungal agents, *Flow Cytometry in Microbiology* (D. Lloyd, ed.), Springer-Verlag, London, 1993, pp. 111–120.
11. T. H. Chrznowski, R. D. Crotty, J. G. Subbard and R. P. Welch, Applicability of the fluorescein diacate method of detecting active bacteria in freshwater, *Microb. Ecol. 10*:179–185 (1984).
12. R. R. Colwell, B. R. Brayton, D. J. Grimes, D. B. Roszak, S. A. Huq, and L. M. Palmer, Viable but non-culturable *Vibrio cholerae* and related pathogens in the envi-

ronment: implications for release of genetically-engineered microorganisms, *Biotechnology 3*:817–820 (1985).

13. J. P. Diaper and C. Edwards, The use of fluorogenic esters to detect viable bacteria by flow cytometry, *J. Appl. Bacteriol. 77*:221–228 (1994).

14. J. P. Diaper and C. Edwards, Survival of *Staphylococcus aureus* in lakewater monitored by flow cytometry. *Microbiology 140*:35–42 (1994).

15. J. P. Diaper and C. Edwards, Flow cytometric detection of viable bacteria in compost, *FEMS Microbiol. Ecol.* (in the press; 1994).

16. J. P. Diaper, K. Tither, and C. Edwards, Rapid assessment of bacterial viability by flow cytometry, *Appl. Microb. Biotechnol. 38*:268–272 (1992).

17. J. Dorsey, C. M. Yentsch, S. Mayer, and C. McKenna, Rapid analytical technique for the assessment of cell metabolic activity in marine microalgae, *Cytometry 10*:622–628 (1989).

18. J. Durodie, K. Coleman, and M. J. Wilkinson, Characterization of bacterial cell size and ploidy using flow cytometry and image analysis, *Flow Cytometry in Microbiology* (D. Lloyd, ed.), Springer-Verlag, London, 1993, pp. 95–109.

19. C. Edwards, The significance of in situ activity on the efficiency of monitoring methods, *Monitoring Genetically Manipulated Microorganisms in the Environment* (C. Edwards, ed.), John Wiley Sons, Chichester, 1993, pp. 1–25.

20. C. Edwards, J. P. Diaper, J. Porter, and R. Pickup, Applications of flow cytometry in bacterial ecology, *Flow Cytometry in Microbiology,* (D. Lloyd ed.), Springer-Verlag, London, 1993, pp. 121–129.

21. C. Edwards, J. Diaper, J. Porter, D. Deere, and R. W. Pickup, Analysis of microbial communities by flow cytometry and molecular probes: identification, culturability and viability, *Beyond the Biomass* (K. Ritz, J. Dighton, and K. E. Giller, eds.), John Wiley & Sons, Chichester, 1994, pp. 57–65.

22. C. Edwards, J. Porter, J. R. Saunders, J. Diaper, J. A. W. Morgan, and R. W. Pickup, Flow cytometry and microbiology, *SGM Q. 19*:105–108 (1992).

23. J. R. Firth, J. P. Diaper, and C. Edwards, Survival and viability of *Vibrio vulnificus* in seawater monitored by flow cytometry, *Lett. Appl. Microbiol. 18*:268–271 (1994).

24. J. C. Gottschal, Phenotypic responses to environmental changes, *FEMS Microbiol. Ecol. 74*:93–102 (1990).

25. J. C. Gottschal, Substrate capturing and growth in various ecosystems, *Appl. Bacteriol. Symp. Suppl. 73*:395–485 (1992).

26. J. L. Jarnagin and D. W. Luchsinger, The use of fluorescein diacetate and ethidium bromide as a stain for evaluating viability of mycobacteria, *Stain Technol. 55*:253–258 (1980).

27. E. S. Kaneshiro, M. A. Wyder, Y.-P. Wu, and M. T. Cushon, Reliability of calcein acetoxy methyl ester and ethidium homodimer or propidium iodide for viability assessment of microbes, *J. Microbiol. Methods 17*:1–16 (1993).

28. A. S. Kaprelyants, and D. B. Kell, Rapid assessment of bacterial viability and vitality using rhodamine 123 and flow cytometry, *J. Appl. Bacteriol. 72*:410–422 (1992).

29. A. S. Kaprelyants, and D. B. Kell, The use of 5-cyano-2,3-ditolyl tetrazolium chloride and flow cytometry for the visualisation of respiratory activity in individual cells of *Micrococcus luteus, J. Microbiol. Methods 17*:115–122 (1993).

30. A. S. Kaprelyants and D. B. Kell, Dormancy in stationary-phase cultures of *Micrococcus luteus*: flow cytometric analysis of starvation and resuscitation, *Appl. Environ. Microbiol. 59*:3187–3196 (1993).

31. A. S. Kaprelyants, J. C. Gottschal and D. B. Kell, Dormancy in non-sporulating bacteria, *FEMS Microbiol. Revs 104*:271–286 (1993).

32. D. B. Kell, H. M. Ryder, A. S. Kaprelyants, and H. V. Westerhoff, Quantifying heterogeneity: flow cytometry of bacterial cultures, Antonie Leeuwenhoek *Microbiol. 60*:145–158 (1991).

33. S. Kjelleberg, K. B. G. Flardh, T. Nystrom, and D. J. W. Moriarty, Growth limitation and starvation of bacteria, *Aquatic Microbiology* (T. E. Ford, ed.), Blackwell, London, 1993, pp. 289–320.

34. S. Kjelleberg, J. Ostling, L. Holmquist, K. Flardh, B. Svenblad, A. Jouper-Jaan, D. Weichart, and N. Albertson, Starvation and recovery of *Vibrio, Trends in Microbial Ecology* (R. Guerrero and C. Pedros-Alio, eds.), Spanish Society for Microbiology, Spain, 1993.

35. R. Kolter, D. A. Siegele, and A. Tormo, The stationary phase of the bacterial cell cycle, *Annu. Rev. Microbiol. 47*:855–874 (1993).

36. J. G. Kramer and F. L. Singleton, Variations in rRNA content of marine *Virbrio* spp. during starvation-survival and recovery, *Appl. Environ. Microbiol. 58*:201–207 (1992).

37. R. Lange and R. Hengge-Aronis, Growth phase-regulated expression of *bol* A and morphology of stationary phase *Escherichia coli* cells are controlled by the novel sigma factor σ^s, *J. Bacteriol, 173*:4474–4481 (1991).

38. H. M. Lappin-Scott and J. W. Costerton, Starvation and penetration of bacteria in soils and rocks, *Experientia 46*:807–812 (1990).

39. D. Lloyd, Flow cytometry: a technique waiting for microbiologists, *Flow Cytometry in Microbiology* (D. Lloyd, ed.), Springer-Verlag, London, 1993, pp. 1–9.

40. B. Lundgren, Fluorescein-diacetate as a stain of metabolically active bacteria in soil, *Oikos 36*:17–22 (1981).

41. A. M. McKay, Viable but non-culturable forms of potentially pathogenic bacteria in water, *Lett. Appl. Microbiol. 14*:129–135 (1992).

42. D. Mason, R. Allman, and D. Lloyd, Uses of membrane potential sensitive dyes with bacteria, *Flow Cytometry in Microbiology* (D. Lloyd, ed.), Springer-Verlag, London, 1993, pp. 67–82.

43. A. Matin, Molecular analysis of the starvation stress in *Escherichia coli. FEMS Microbiol. Ecol. 74*:185–196 (1990).

44. T. Matsuyama, Staining of living bacteria with rhodamine 123, *FEMS Microbiol. Lett. 21*:153–157 (1984).

45. J. J. Mekalonos, Environmental signals controlling the expression of virulence determinants in bacteria, *J. Bacteriol. 174*:1–7 (1992).

46. R. Y. Morita, Starvation and miniaturisation of heterotrophs with special emphasis on maintenance of the starved viable state, *Bacteria in Their Natural Environments* (M. M. Fletcher and G. D. Floodgate, eds.), Academic Press, London, 1985, pp. 111–130.

47. R. Y. Morita, The starvation-survival state of microorganisms in nature and its relationship to the bioavailable energy, *Experientia 46*:813–817 (1990).

48. A. E. Musgrave and D. W. Edley, Measurement of intracellular pH, *Methods in Cell Biology,* Vol. 33, (Z. Darzynkiewicz and H. A. Crissman, eds.), Academic Press, New York, 1990, pp. 56–69.

49. R. Nir, Y. Yisraeli, R. Lamed, and E. Sahar, Flow cytometric sorting of viable bacteria and yeasts according to ß-galactosidase activity, *Appl. Environ. Microbiol. 56*:3861–3866 (1990).

50. J. A. Novitsky and R. Y. Morita, Survival of a psychrophilic marine *Vibrio* under long-term nutrient starvation, *Appl. Environ. Microbiol. 33*:635–641 (1977).

51. T. Nystrom, N. H. Albertson, K. Flardh, and S. Kjelleberg, Physiological and molecular adaptation to starvation and recovery from starvation by the marine *Vibrio* sp. S14, *FEMS Microbiol. Ecol. 74*:129–140 (1990).

52. T. Nystrom, R. M. Olsson, and S. Kjelleberg, Survival, stress resistance, and alterations in protein expression in the marine *Vibrio* strain S14 during starvation for different individual nutrients, *Appl. Environ. Microbiol. 58*:55–65 (1992).

53. J. D. Oliver and D. Wanucha, Survival of *Vibrio vulnificus* at reduced temperature and elevated nutrient, *Food Safety 10*:79–86 (1989).

54. J. D. Oliver, R. A. Warner, and D. R. Cleland, Distribution of *Vibrio vulnificus* and other lactose-fermenting vibrios in the marine environment, *Appl. Environ. Microbiol 45*:985–998 (1983).

55. J. D. Oliver, L. Nilsson, and S. Kjelleberg, Formation of nonculturable *Vibrio vulnificus* cells and its relationship to the starvation state, *Appl. Environ. Microbiol. 57*:2640–2644 (1991).

56. K. R. O'Neil, S. H. Jones, and D. J. Grimes, Seasonal incidence of *Vibrio vulnificus* in the Great Bay estuary of New Hampshire and Maine, *Appl. Environ. Microbiol. 58*:3257–3262 (1992).

57. R. W. Pickup, Development of methods for the detection of specific bacteria in the environment, *J. Gen. Microbiol. 137*:1009–1019 (1991).

58. A. C. Pinder, C. Edwards, R. G. Clarke, J. P. Diaper, and S. A. G. Poulter, Detection and enumeration of viable bacteria by flow cytometry, *New Techniques in Food and Beverage Microbiology* (R. G. Kroll, A. Filmour, and M. Sussman, eds.), SAB Technical Series No. 31, Blackwell, London, 1993, pp. 67–86.

59. G. G. Rodriquez, D. Phipps, K. Ishiguro, and H. F. Ridgeway, Use of a fluorescent redox probe for direct visualization of actively respiring bacteria, *Appl. Environ. Microbiol. 58*:1801–1808 (1993).

60. D. B. Roszak and R. R. Colwell, Metabolic activity of bacterial cells enumerated by direct viable count, *Appl. Environ. Microbiol. 53*:2889–2983 (1987).

61. M. Schaechter, Going after the growth curve, *The Molecular Biology of Bacterial Growth* (M. Schaechter, F. C. Niedhardt, J. L. Ingraham, and N. O. Kjeldgaard, eds.), Jones & Bartlett, Boston, 1985, pp. 370–372.

62. D. A. Siegele and R. Kolter, Life after log, *J. Bacteriol. 174*:345–348 (1992).

63. M. P. Spector, Gene expression in response to multiple nutrient-starvation conditions in *Salmonella typhimurium, FEMS Microbiol. Ecol. 74*:175–184 (1990).

64. B. K. Thorsen, O. Enger, S. Norland and K. A. Hoff, Long term starvation of *Yersinia ruckeri* at different salinities studied by microscopical and flow cytometric methods, *Appl. Environ. Microbiol. 58*:1624–1628 (1992).

65. F. Torrella, and R. Y. Morita, Microcultural study of bacterial size changes and microcolony and ultramicrocolony formation by heterotrophic bacteria in sea water, *Appl. Environ. Microbiol. 41*:518-527 (1981).

66. R. A. Van Bogelen and F. C. Neidhardt, Global systems approach to bacterial physiology: protein responders to stress and starvation, *FEMS Microbiol. Ecol. 74*:121–128 (1990).

67. D. Weichart, J. D. Oliver, and S. Kjelleberg, Low temperature induced non-culturability and killing of *Vibrio vulnificus. FEMS Microbiol. Lett. 100*:205–210 (1992).

68. M. M. Zambrano, D. A. Siegele, M. Almiron, A. Tormo, and R. Kilter, Microbial competition: *Escherichia coli* mutants that take over stationary phase cultures, *Science 259*:1757–1760 (1993).

17

Advances in the Flow Cytometric Characterization of Plant Cells and Tissues

DAVID W. GALBRAITH AND GEORGINA M. LAMBERT
University of Arizona, Tucson, Arizona

I. INTRODUCTION

Flow cytometry has increasingly found important applications in the study of higher plants, their cells, and organelles. Plants have provided some of the most intriguing challenges to the routine application of this technology. This is partly because higher plants are not suspensions of single cells, but typically comprise complex three-dimensional tissues. For plant tissues to be studied through flow cytometry, suspensions of single cells or organelles must be prepared before analysis. The act of preparing plant cells for flow cytometry involves perturbation of their normal environment, and this can alter the parameters measured through flow cytometry.

Applications of flow cytometry to plant systems have also been limited because there has been considerably less investment into the cell biology of plants, compared with that in animals. For example, for the mammalian biologist, a large number of cell surface markers have been identified. Poly- and monoclonal antibodies recognizing these markers are typically available, and often the genes encoding these markers have been identified and characterized.

This provides an enormous resource for a wide variety of applications involving flow cytometry of mammalian cells; such a resource does not exist for plant cell biologists.

In the first part of this chapter, we outline some general methods for the analysis of plant cells and protoplasts using flow cytometry. We go on to discuss methods for the analysis of subcellular organelles, specifically focusing on analysis of nuclear DNA contents. In the second part, we discuss problems specific to the sorting of plant protoplasts and organelles, concentrating on large particles, fragility, and sort purity.

II. MATERIALS AND METHODS

A. Plant Materials

Tobacco (*Nicotiana tabacum* L. cv. Xanthi) plants were maintained as axenic shoot cultures at 22°C under continuous light, as previously described [1]. Arabidopsis plants (*Arabidopsis thaliana* L. ecotype Columbia) were grown under sterile conditions, as previously described [2]. Plants raised under nonsterile conditions [3] were maintained under an 8-hour light cycle at 22°C within standard growth chambers (Conviron).

B. Chemicals

Macerase, cellulysin, and propidium iodide (PI) were obtained from Calbiochem, Inc. (La Jolla, CA), and fluorescent microspheres from the Coulter Corporation (Miami, FL). MS (Murashige and Skoog) medium was from Gibco (Grand Island, NY). All remaining chemicals were obtained from the Sigma Chemical Co. (Saint Louis, MO).

C. Protoplast Preparation

All procedures were carried out using standard sterile techniques. Fully expanded leaves were excised from axenic tobacco or arabidopsis plants (approximately 600-mg wet weight). They were sliced in 20 ml of digestion medium, contained in a sterile plastic petri dish. The digestion medium comprised 0.1% Driselase, 0.1% macerase, and 0.1% cellulysin, dissolved in a buffer containing 0.5 M mannitol, 10 mM $CaCl_2$, and 3 mM MES, pH 5.7, and was sterilized by Millipore filtration (GSWP 047). Incubation was continued at 22°C overnight (18–20 h) in darkness without agitation. The suspension of protoplasts was then filtered through two layers of sterile cheesecloth into a 50-ml sterile centrifuge tube, and was centrifuged at 50 × g for 8 min. The pelleted protoplasts were gently resuspended in 20 ml of 25% (w/v) sucrose (dissolved in modified To medium; [1]), overlaid with 5 ml of W5 medium [4], and centrifuged at 50 × g for 10 min. Viable protoplasts, which accumulated at the interface, were carefully removed using a wide-bore Pasteur pipette, and were diluted by addition of two volumes of W5 medium. Protoplast numbers were determined using a Fuchs-Rosenthal hemocytometer.

D. Preparation of Subcellular Homogenates

Whole tobacco or arabidopsis leaves or maize kernels (200–500 mg) were transferred into a plastic petri dish (60 mm diameter), placed on a prechilled (4°C) ceramic tile. Ice-cold chopping buffer (3 ml) was added, and the tissues were chopped for 1.5 min using a disposable, single-edged razor blade (VWR, Phoenix, AZ). The chopping buffer (CB) comprised 45mM MgCl$_2$, 30 mM sodium citrate, 20 mM MOPS, and 0.01–1% (w/v) Triton X-100, adjusted to pH 7.0 using 0.1 M NaOH. For arabidopsis, the buffer was modified by addition of 5% (w/v) polyvinylpyrrolidone, 10 mM ascorbic acid, and 10 mM dithiothreitol. The resultant homogenates were filtered through 15-μm nylon mesh (Tetko, Briarcliff Manor, NY). These homogenates yielded intact nuclei suitable for flow analysis and sorting [5].

E. Staining of Nuclei

DNase-free ribonuclease (final concentration 0.1 mg/ml) was added, followed by PI (final concentration of 100 μg/ml; dispensed from a 2.5-mg/ml stock solution made up in H$_2$O and stored in darkness at room temperature). The homogenates were maintained on ice. Flow analysis was usually performed 15–30 min after addition of the PI and ribonuclease.

F. Flow Analysis and Sorting

All analyses and sorting were done using a Coulter Elite flow cytometer equipped with a 20-mW argon laser (488-nm emission), a forward-angle light-scatter (FALS) detector, and four photomultipliers (PMTs). The standard filter configuration assigned 90°-light scatter to PMT1, 505 to 545-nm fluorescence (green) to PMT2, 555 to 595-nm flourescence (orange) to PMT3, and 670 to 680-nm (red) fluorescence to PMT4. Two sort–sense flow cells were employed in this work, one having a 100-μm and the other a 150-μm orifice. All sheath and sample fluids, with the exception of deionized water, were filtered through 0.22-μm Millipore GSWP 047 filters before use. The cytometer was aligned using calibration fluorospheres (DNA Check, Coulter Electronics, Hialeah, FL) diluted by addition of 10 volumes of deionized water. Uniparametric histograms of fluorescence emission (integral signal) and FALS were collected at a sample flow rate of 70 particles per second, with the FALS discriminator being set to 100, and all other discriminators being switched off. The voltages and amplification settings for the various applications are given in Table 1. Adjustment of

TABLE 1 Voltage and Amplification Settings Used for the Specified Applications of the Flow Cytometer

Application	Discrim-inator	HV/AMP			
		PMT1	PMT2	PMT3	PMT4
Fluoro-spheres	FALS (100)	380 2.0 (I)[a]	570 7.5 (I) 7.5 (P)[a]	550 7.5 (I) 7.5 (P)	650 7.5 (I) 7.5 (P)
Pollen	FALS (100)	140 7.5	650 7.5 (I) 1.0 (P)		
Proto plasts	FALS (100)	400 2.0			580 10.0 (I) 7.5 (P)
Nuclei/ SCH[c]	PMT3 (30; I)	350 2.0	580 7.5 (I) 7.5 (P)	700–1000[b] 7.5 (I) 5.0 (P)	660 7.5 (I) 7.5 (P)

[a]I, integral signal; P, peak signal.
[b]Dependent on genome size.
[c]SCH, subcellular homogenates.

the optics was continued until population coefficients of variation (CVs) for pulse integral and FALS were minimized; these were typically less than 2% for fluorescence and 2–3% for FALS.

1. Analysis of Plant Cells, Protoplasts, and Subcellular Organelles

Typical instrument settings for the various applications are given in Table 1. In all cases the laser was operated at 15 mW, and the standard optical filter configuration was used.

Pollen

Pollen is autofluorescent, so signals from PMT2 can be accumulated without the need for staining the cells. We generally accumulate histograms of FALS, 90°-light scatter, and integral green fluorescence (PMT2; both linear and log signals). The sheath fluid is water or phosphate-buffered saline (PBS).

Protoplasts

For analysis of unlabeled leaf protoplasts, autofluorescence from chlorophyll provides the major analytical signal. We generally accumulate histograms of

FALS, 90°-light scatter, and integral red fluorescence (PMT4; both linear and log signals). The sheath fluid is W5 medium [1].

Subcellular Homogenates

For subcellular homogenates, CB is employed as the sheath fluid. For analysis of PI-stained nuclei, the Elite is triggered on orange fluorescence (PMT3), with the discriminator (integral signal) set to 30. Integral signals are collected on both linear and log scales. The HV settings are adjusted according to the amounts of DNA present in the nuclei so that the signals fall within the dynamic range of the amplifiers. In all cases, two-dimensional histograms of time versus fluorescence are routinely accumulated as controls; staining intensities should be constant over time. If nonlinearities are observed, the results should be discarded.

We find that about ten samples per hour can be conveniently processed for flow analysis using the chopping procedure.

2. Sorting of Plant Cells and Subcellular Particles

General Considerations

For routine analysis of small particles and cells using the 100-µm flow tip, the sheath and sample pressures are set to 12 and 11.5 psi, respectively, and the electromechanical transducer is adjusted to about 16.7 kHz. When sorting larger (> 10-µm) particles and cells, we lower the sheath and sample pressures to 8.0 and 7.5 psi respectively; operation of the electromechanical transducer at 70% amplitude then gives a stable sorted stream within the range of 12–17 kHz. Lower-drive frequencies improve the efficiency of recovery of larger particles [6]. For the 150-µm flow tip, a drive amplitude of 50% and a frequency of 11.5 kHz gave stable and efficient sorting of large particles (pecan pollen).

Optimizing the sorting process involves the following steps. The transducer is left to warm-up through operation at 70% amplitude for 1 h. The droplet deflection assembly is then moved upward to a point as close as possible to the flow cell tip, without blocking the laser beam. The view of the flow stream with the videocamera is adjusted so that the ground plate of the deflection assembly can be observed on the left edge of the screen and the laser beam intercept on the right using only the "pan" function. The transducer drive frequency and amplitude are then adjusted to provide the shortest possible droplet break-off point. After switching to sort test mode, the delay setting is adjusted in 0.1-drop increments ("phase" adjustment) to give a stable, sorted stream. The deflection plate assembly is then lowered so that two to three free droplets can be seen above the ground plane. The fluid stream is then observed using the videocamera. The last-attached droplet should exhibit a well-rounded profile on the left-hand (nonattached) side, and should be obviously connected by a ligament to the flow stream on the right-hand side. If smaller satellite droplets are observed, it should be determined whether these are "fast" or "slow," defined as to whether they merge with the major droplet ahead of or behind the satellite. The amplitudes and frequencies of transducer activation should be adjusted until only fast satellites are produced. These carry charge of the same sign as the droplets with which they merge and, therefore, do not affect their electrostatic deflection.

The cursor is next moved to superimpose the second well-defined undulation to the right of the last-attached droplet. With the cursor, a second point is marked between the first two free drops above the ground plane. The number of droplets separating the two cursors is entered into the delay-calculation program. The resultant delay setting should be very close to the delay setting defined later as optimal through empirical sorting of particles. Histograms are now acquired, using standard particles that, ideally, are close to the size of the cells of interest [7], and sort windows are programmed based on these distributions. A sort matrix analysis is done, to empirically define a delay setting that maximizes sort efficiencies (near-100% values should be achieved). This is measured through batch sorting of groups of 25 particles onto standard 3 × 1" glass microscope slides, and counting the numbers actually recovered under the microscope. Between each batch of 25, the sort delay setting is adjusted by 1.0-step increments to span a range of ± 5.0 around the calculated delay setting. For one-drop sorts, the delay is adjusted in 0.1-drop increments. It should be emphasized that sorting efficiencies for large, fragile particles (such as protoplasts) should be optimized using indestructible particles of equivalent sizes, such as pollen [6]. The next step is to adjust the deflection plate high-voltage and the phase settings to obtain side streams without droplet "fanning." Histograms are then accumulated for the particles or cells of interest, the appropriate sort windows are defined, and sorting is enabled.

For the 150-μm tip, the point of droplet formation is usually below the viewing field of the videocamera. This means that the cursor-based computation just described cannot be done. Here, the sort delay has to be determined empirically (i.e. by performing a sort matrix).

Sterile Sorting

When protoplasts were to be cultured after sorting, sterile sorting procedures were as follows. Sheath and rinse tanks were cleaned with detergent, then filled with 70% ethanol. Ethanol was back-flushed through the sample uptake module, then a sample tube containing 70% ethanol was run through the system for 15 min. These steps were repeated with sterile (autoclaved) water to rinse the system before filling with sterile sheath fluid. Samples were sorted into sterile 15-ml centrifuge or microfuge tubes.

Three-Droplet and One-Droplet Sorting

For the Elite, the user is able to define the numbers of droplets that are charged for each event that is sorted. In standard, three-droplet, sort mode, with a transducer drive frequency of 16.7 kHz and a sort delay of 15, we routinely obtained sort efficiencies of 100%. Three-droplet sorting provides the greatest margin of error in terms of recovery of desired particles or cells. However, it increases the probability that unwanted particles will be included within the sorted droplets. As long as these unwanted particles trigger data acquisition, they can be eliminated by enabling the anticoincidence circuit, although this may yield unacceptably low recoveries. If the unwanted particles do not trigger data acquisition, typically because they produce signals lower than the threshold set on the active discriminator, they are in essence invisible to the flow cytometer. The only means to reduce or eliminate their presence is to perform one-droplet sorts. Since the error margin for the desired particle is reduced in one-droplet sorting, careful adjustment to the phase setting is required to optimize sort recoveries.

Sequential Sorting

For certain applications, for example, sorting of G_1 and G_2 nuclei from cellular homogenates, sequential sorts are required to achieve satisfactory levels of purity (> 95%). The only modification required for this involves maintaining the sample input and collection tubes on ice. This can be conveniently done using ice contained in small, thin-walled plastic bags.

For sequential sorting of nuclei, we typically employed the one-droplet sort mode at a frequency of 16.7 kHz, a drive amplitude of 70%, and a delay setting of about 15. The discriminators were set to trigger on the orange fluorescence signal produced by PI-stained nuclei (PMT3), and two-dimensional histograms of peak versus integral signals were acquired. A bit-mapped region corresponding to the G_2 nuclei was used as the sort region. The sample flow rate was adjusted to give a data rate of approximately 300–500 particles per second for the first round of sorting. The anticoincidence circuit was enabled. The sorted nuclei were then restained with propidium iodide, reanalyzed, and resorted at a data rate of about 100 particles per second, with the anticoincidence circuit again enabled.

III. RESULTS AND DISCUSSION

A. Flow Cytometry

1. General Considerations

Because in flow cytometry, the arrival of particles at the detection area is a stochastic event, the analysis circuits of flow cytometers are designed so that each arrival triggers data acquisition from that particle. Triggering is typically based on the FALS signal produced by the particle, is enabled through activation of the relevant discriminator, and requires adjustment of the level of this discriminator to eliminate spurious triggering by noise. When a single type of

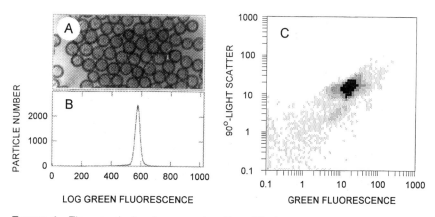

FIGURE 1 Flow analysis of ragweed pollen: (A) Appearance of the population under the light microscope, magnification × 330; (B) uniparametric analysis of green fluorescence emission, employing logarithmic amplification; (C) biparametric analysis of 90°-light scatter versus green fluorescence, employing logarithmic amplification.

cell or particle predominates within a suspension, it is simple to set up the appropriate flow cytometric analysis. For complex suspensions of biological cells and subcellular homogenates, observation and identification of the particle population of interest may become difficult, especially if it comprises a relatively minor proportion of the total particles present in the suspension. This problem is detailed for the different types of analyses that are presented.

2. Flow Analysis of Pollen

Analysis of ragweed pollen is relatively straightforward, since suspensions contain little debris (Fig. 1A), and the discriminator setting (using FALS) can be readily adjusted to exclude noise. A one-dimensional histogram reveals a single well-defined class of fluorescent particles within the total population (see Fig. 1B). Some substructure to the populations is suggested by two-dimensional analysis (see Fig. 1C), but integration indicates that 98% of the population falls within the major class.

3. Flow Analysis of Protoplasts According to Chlorophyll Content

A typical preparation of leaf protoplasts of *Nicotiana tabacum* and the corresponding uniparametric and biparametric flow analyses are given in Figure 2. Although the protoplast population appears pure under the microscope (see Fig. 2A), uniparametric analysis shows a bimodal distribution (see Fig. 2B). The intact protoplasts comprise a peak located at about channel 700. Chloroplasts released from broken protoplasts are found as a broad peak of lower mean fluo-

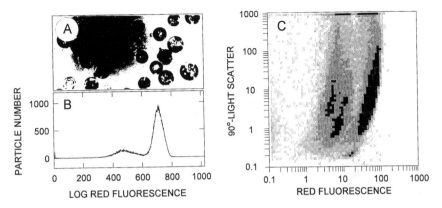

FIGURE 2 Flow analysis of tobacco leaf protoplasts: (A) Appearance of the protoplasts under the light microscope, magnification × 240; (B) uniparametric analysis of red fluorescence emission, employing logarithmic amplification; (C) biparametric analysis of 90°-light scatter versus red fluorescence, employing logarithmic amplification.

rescence (located from channels 350–600). Since individual protoplasts contain large numbers of chloroplasts [8], the breakage of a minor proportion of the protoplasts is responsible for the presence of the chloroplasts within the frequency distributions. Two-dimensional analysis of the populations based on red fluorescence and 90°-light scatter reveals substructure to the chloroplast population (see Fig. 2C), but the basis for this substructure is not yet understood. In these analyses, again, triggering is based on FALS, since the components of interest in the protoplast suspension (the protoplasts and chloroplasts) both scatter light and exhibit fluorescence, and there are few nonfluorescent light-scattering particles present.

4. Flow Analysis of Subcellular Organelles Within Cell Homogenates

In contrast with the situation described earlier, when intact plant tissues are homogenized and the homogenates are used for flow cytometric analysis, a considerable proportion of the homogenate comprises nonfluorescent, light-scattering particles. These are derived from many of the different organelles within the cytoplasm, as well as from subcellular debris. As a consequence, if flow analysis is triggered on FALS, homogenates appear as nondescript distributions in which the frequencies decrease exponentially as one moves across the abscissa from channel 1. Data acquisition rates appear anomalously high, and the populations of interest (for example, PI-stained nuclei, which comprise a very minor proportion of the total light-scattering population) may be completely hidden. This situation can be avoided by triggering on the fluorescent signal of interest;

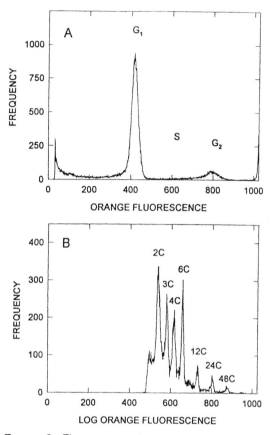

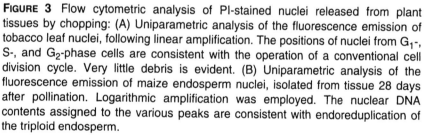

FIGURE 3 Flow cytometric analysis of Pl-stained nuclei released from plant tissues by chopping: (A) Uniparametric analysis of the fluorescence emission of tobacco leaf nuclei, following linear amplification. The positions of nuclei from G_1-, S-, and G_2-phase cells are consistent with the operation of a conventional cell division cycle. Very little debris is evident. (B) Uniparametric analysis of the fluorescence emission of maize endosperm nuclei, isolated from tissue 28 days after pollination. Logarithmic amplification was employed. The nuclear DNA contents assigned to the various peaks are consistent with endoreduplication of the triploid endosperm.

for most of our applications, involving cell cycle analyses, this is done using the peak signal from PMT3, and activating the discriminator for this channel only, either on the peak or integral signal, depending on which is to be collected in the form of histograms.

The Cell Division Cycle

When stained with a DNA-specific fluorochrome, nuclei released during homogenization may be analyzed for DNA content, ploidy, and cell cycle status. The standard laser configuration in the Elite (argon or HeNe illumination) restricts our choice of fluorochromes to PI and (perhaps) TOTO and YOYO [9]. For tobacco leaves, the chopping process releases nuclei from cells exclusively within a conventional diploid cell cycle, in which the DNA varies from 2C to 4C (Fig. 3A). For typical leaves, most nuclei (in this case, about 70%) are from cells in G_0/G_1 and have a 2C DNA content of 9.2 pg [5].

For developing maize kernels, which lack mature chloroplasts, the chopping process releases nuclei of varying degrees of endoreduplication (see Fig. 3B). Development of maize kernels follows a process of double fertilization, in which two haploid sperm cells fuse with one diploid and one haploid egg cell, respectively. This gives rise to the triploid endosperm tissue and to the diploid embryo, The nuclei released by chopping the kernels are thus derived from maternal (diploid) cells, progeny (diploid) cells, and endosperm (triploid) cells. Analysis of the positions of the various peaks (see Fig. 3B) indicates that the triploid cells are highly endoreduplicated (peaks in an arthmetic series reaching 48C peaks can be readily discerned), whereas the diploid cells are not. In this display, the use of a logarithmic scale is essential to the display of the various peaks, since the range from lowest to highest greatly exceeds the range available under linear amplification. We have reported systemic endoreduplication in many different types of plants, including succulents [10] and arabidopsis [2].

Nuclear–Cytoplasmic Interactions

When green tissues are chopped, the presence of chloroplasts greatly increases the complexity of multiparametric flow histograms. In Figure 4, we illustrate the biparametric flow analysis of PI-stained homogenates of *A. thaliana*. Since the fluorescence emission spectra of chloroplasts and PI-stained nuclei are different, they form separate clusters; chloroplasts comprise a single discrete population, whereas the endoreduplicated nuclei are found in three small clusters, falling on a correlated line (see Fig. 4A). Information about chloroplast size and number can be obtained by measurement of the intensity of the red fluorescence signal and by integrating the numbers found within the different areas. In Figure 4B–D, we illustrate this through two-parameter flow analysis of three *arc* mutants of *A. thaliana*. These mutants possess defects in the regulation of normal chloroplast replication. In the cases of *arc2* and, particularly, *arc3*, the numbers of chloroplasts per cell are drastically reduced, and their individual sizes (hence, fluorescence emission) are increased [11]. This is reflected by an increase in total fluorescence per chloroplast (see Fig. 4C,D). The fact that the chopping technique can be employed to screen large numbers of individual plants argues that flow

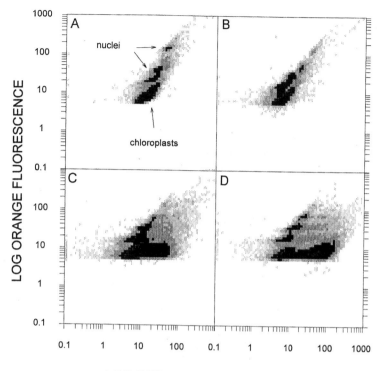

FIGURE 4 Biparametric analysis of PI-stained *Arabidopsis thaliana* leaf homogenates. Two distinct clusters are apparent, corresponding to the nuclei and chloroplasts. (A) Wild-type; (B) *arc1* mutant; (C) *arc2* mutant; (D) *arc3* mutant.

cytometry is a reasonable way to rapidly identify and characterize further mutants defective in chloroplast replication.

B. Cell Sorting

When applying cell sorting techniques to higher plants, two problems are frequently encountered: (a) optimizing sort recoveries with large flow cell tips, and (b) optimizing sort purities.

1. Large Particle Sorting

Correct piezoelectric actuation of flow tips larger than those conventionally found on jet-in-air sorters (60–80 μm) is governed by theoretical considerations, in particular that, for droplet formation to occur, the wavelength of the imposed undulation must be greater than the diameter of the fluid jet multiplied by π [7].

The position of droplet break-off is also a direct function of the rate of expulsion of liquid from the flow cell tip, which is a function of the sheath pressure applied to the flow tip. The use of the 150-μm tip is less convenient than that of the 100-μm tip, since the droplet break-off point occurs below the viewing area; modifications to the videocamera configuration would resolve this problem, but would appear beyond the capabilities of most flow facilities.

The requirements for stable sorting of large particles using large flow cell tips have been comprehensively discussed elsewhere [6,7]. Here we would simply stress the importance of employing sheath pressures and piezoelectric-drive frequencies that are lower (often much lower) than those typically employed for animal cell sorting. We also stress the importance of optimizing the sort process using indestructible particles that have diameters similar to the fragile cells that are to be subsequently sorted (pollen conveniently fulfills this role). Large particles have a tendency to perturb the process of droplet formation, and sort conditions that appear optimal for small particles frequently are nonoptimal for larger cells [6].

Under the conditions described in the methods section (Sec. II.F.2), pecan pollen is recovered at efficiencies of 100% using either the 100-μm or 150-μm flow tips. Sorted protoplasts are recovered at lower efficiencies (from 50-100% depending on sample), even if pollen recoveries are 100%. The ability to recover viable protoplasts after sorting is governed almost exclusively by the "quality" of the protoplasts before sorting. Inspection under the light microscope should reveal spherical cells within which the chloroplasts are uniformly distributed (see Fig. 2). The protoplasts should have a smooth, intact plasma membrane, and there should be minimal subcellular debris.

2. Sort Purity

Many applications of cell sorting involve placing the sorted cells into culture to increase the amounts of materials for study. With the exception of rare-event sorting, it is relatively easy to establish conditions under which specific subpopulations of cells can be highly enriched through cell sorting. This is because triggering on cells is unambiguous, and because the presence of noncellular debris on sorted cells rapidly becomes irrelevant as they grow. The problem of ensuring high levels of sort purity becomes acute, however, when the sorted material is to be subjected to immediate biochemical or molecular analysis, when the contaminants are particularly abundant, and when they do not trigger detection by the flow cytometer. This is illustrated by the use of two sequential cycles of one-drop sorting to yield purified populations of G_2 nuclei from tobacco homogenates.

In Figure 5A, we show a uniparametric flow analysis of PI-stained *N. tabacum* leaf nuclei. The G_1 nuclei appear in channels 206–230 and the G_2 nuclei in channels 396–445. Biparametric analysis of peak versus integral fluo-

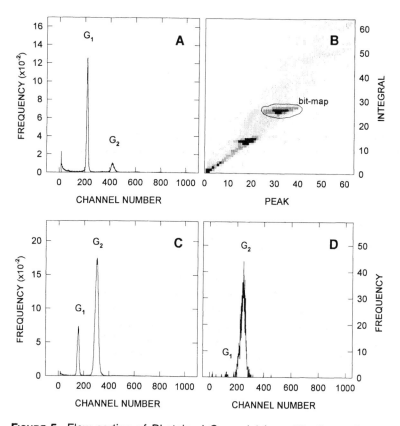

FIGURE 5 Flow sorting of PI-stained G_2 nuclei from *Nicotiana tabacum* leaves. (A) Uniparametric analysis of the starting population; (B) biparametric analysis of the starting population, plotting the peak versus integral signal of PI-dependent fluorescence. The bit-map employed for sorting is indicated. (C) Uniparametric analysis of the nuclei, after one round of sorting; (D) uniparametric analysis of the nuclei, after two rounds of sorting.

rescence (see Fig. 5B) allows positioning of amorphous bit-maps for sorting; here, it includes only the G_2 nuclei. Following one round of sorting, the proportions of G_2 nuclei are considerably enhanced (see Fig. 5C). The major contaminant comprises G_1 nuclei. This is probably due to the breaking apart of adhered G_1 doublets during sorting; inspection of Figure 5B reveals the presence of minor clusters of nuclei corresponding to three and four times the 2c DNA content, which implies that the 4c peak will include some G_1 doublets. A second round of sorting eliminates almost all of the contaminants, yielding a pure population of G_2 nuclei (see Fig. 5D).

When sorting organelles from cellular homogenates, one frequently encounters contamination from other particles present in the homogenate. In homogenates prepared from whole leaves, free chloroplasts are a potential problem because they constitute most of the total organellar mass. Inclusion of Triton-X 100 in the chopping buffer results in the dissolution of the chloroplast membranes, but does not affect the integrity of the nuclear membrane. This essentially eliminates problems of chloroplast contamination for this particular application. However, in other cases, inclusion of detergent may not be permissible; dilution of the sample and reduction of the sample flow rate should minimize chloroplast contamination in this situation.

IV. FUTURE PROSPECTS

The unique capabilities of flow cytometry and cell sorting, to provide data concerning the optical properties of cells and to use this for the selective purification of different types of cell or particles, have yet to be fully explored with higher plant systems. Recently, developed techniques of molecular biology provide considerable impetus to this exploration. These developments include the transgenic expression of the green fluorescent protein of *Aequorea victoriae* within eukaryotic cells [12]. Fluorescent tagging of cells according to gene expression, and the subsequent sorting of these cells based on GFP fluorescence would afford a powerful new means for analyzing the regulation of gene expression in multicellular organisms. Other recently developed methods of differential display [13] offer the ability to identify and clone genes that are expressed within small amounts of tissues; such as, those in cells isolated by cell sorting. As our quest to analyze eukaryotic development focuses more finely on specific cell types, which often are limited in numbers, the importance of flow cytometry and cell sorting in the isolation of these cell types will become increasingly evident.

ACKNOWLEDGMENTS

We thank Dr. Gideon Grafi for providing us with the maize kernels used in the experiment described in Figure 3B. This work was supported in part by the National Science Foundation and the Competitive Grants Program of the United States Department of Agriculture.

REFERENCES

1. K. R. Harkins, R. A. Jefferson, T. A. Kavanaugh, M. W. Bevan, and D. W. Galbraith, Expression of photosynthesis-related gene fusions is restricted by cell type in transgenic plants and transfected protoplasts, *Proc. Natl. Acad. Sci. USA* 87:816 (1990).

2. D. W. Galbraith, K. R. Harkins, and S. Knapp, Systematic endopolyploidy in *Arabidopsis thaliana, Plant Physiol. 96*:985 (1991).
3. N. R. Forsthoefel, Y. Wu, B. Schulz, M. J. Bennett, and K. A. Feldmann, T-DNA insertion mutagenesis in *Arabidopsis*: propects and perspectives, *Aust. J. Plant Physiol. 19*:353 (1992).
4. I. Negrutiu, R. Shillito, I. Potrykus, G. Biasini, and F. Sala, Hybrid genes in the analysis of transformation conditions, *Plant Mol. Biol. 15*:363 (1987).
5. G. Bharathan, G. Lambert, and D. W. Galbraith, Nuclear DNA content of mono-cotyledons and related taxa, *Am. J. Bot. 81*:381 (1994).
6. K. R. Harkins and D. W. Galbraith, Factors governing the flow cytometric analysis and sorting of large biological particles, *Cytometry 8*:60 (1987).
7. D. W. Galbraith, Large particle sorting, *Flow Cytometry and Cell Sorting* (A. Radbruch, ed.), Springer-Verlag, Berlin, 1992, pp. 189–204.
8. K. R. Harkins and D. W. Galbraith, Flow sorting and culture of plant protoplasts, *Physiol. Plant 60*:43 (1984).
9. G. T. Hirons, J. J. Fawcett, and H. A. Crissman, TOTO and YOYO: new very bright fluorochromes for DNA content analysis by flow cytometry, *Cytometry 15*:129 (1994).
10. E. J. DeRocher, K. R. Harkins, D. W. Galbraith, and H. J. Bohnert, Developmen-tally-regulated systemic endopolyploidy in succulents with small genomes, *Science 250*:99 (1990).
11. K. A. Pyke and R. M. Leach, Chloroplast division and expansion is radically altered by nuclear mutations in *Arabidopsis thaliana, Plant Physiol. 99*:75 (1988).
12. M. Chalfie, Y. Tu, G. Euskirchen, W. Ward, and D. C. Prasher, Green fluorescent protein as a marker for gene expression, *Science 263*:802 (1994)
13. P. Liang and A. B. Pardee, Differential display of eukaryotic messenger RNA by means of the polymerase chain reaction, *Science 257*:967 (1992).

Index